UCLA Symposia on Molecular and Cellular Biology, New Series

Series Editor, C. Fred Fox

RECENT TITLES

Volume 56
Cellular and Molecular Biology of Tumors and Potential Clinical Applications, John Minna and W. Michael Kuehl, *Editors*

Volume 57
Proteases in Biological Control and Biotechnology, Dennis D. Cunningham and George L. Long, *Editors*

Volume 58
Growth Factors, Tumor Promoters, and Cancer Genes, Nancy H. Colburn, Harold L. Moses, and Eric J. Stanbridge, *Editors*

Volume 59
Chronic Lymphocytic Leukemia: Recent Progress and Future Direction, Robert Peter Gale and Kanti R. Rai, *Editors*

Volume 60
Molecular Paradigms for Eradicating Helminthic Parasites, Austin J. MacInnis, *Editor*

Volume 61
Recent Advances in Leukemia and Lymphoma, Robert Peter Gale and David W. Golde, *Editors*

Volume 62
Plant Gene Systems and Their Biology, Joe L. Key and Lee McIntosh, *Editors*

Volume 63
Plant Membranes: Structure, Function, Biogenesis, Christopher Leaver and Heven Sze, *Editors*

Volume 64
Bacteria–Host Cell Interaction, Marcus A. Horwitz, *Editor*

Volume 65
The Pharmacology and Toxicology of Proteins, John S. Holcenberg and Jeffrey L. Winkelhake, *Editors*

Volume 66
Molecular Biology of Invertebrate Development, John D. O'Connor, *Editor*

Volume 67
Mechanisms of Control of Gene Expression, Bryan Cullen, L. Patrick Gage, M.A.Q. Siddiqui, Anna Marie Skalka, and Herbert Weissbach, *Editors*

Volume 68
Protein Purification: Micro to Macro, Richard Burgess, *Editor*

Volume 69
Protein Structure, Folding, and Design 2, Dale L. Oxender, *Editor*

Volume 70
Hepadna Viruses, William Robinson, Katsuro Koike, and Hans Will, *Editors*

Volume 71
Human Retroviruses, Cancer, and AIDS: Approaches to Prevention and Therapy, Dani Bolognesi, *Editor*

Volume 72
Molecular Biology of the Human Brain, Edward G. Jones, *Editor*

Volume 73
The T-Cell Receptor, Mark M. Davis and John Kappler, *Editors*

Volume 74
Growth Regulation of Cancer, Marc E. Lippman, *Editor*

Volume 75
Steroid Hormone Action, Gordon Ringold, *Editor*

Volume 76
Molecular Biology of Intracellular Protein Sorting and Organelle Assembly, Ralph A. Bradshaw, Lee McAlister-Henn, and Michael G. Douglas, *Editors*

Volume 77
Signal Transduction in Cytoplasmic Organization and Cell Motility, Peter Satir, John Condeelis, and Elias Lazarides, *Editors*

Volume 78
Tumor Progression and Metastasis, Garth L. Nicolson and Isaiah J. Fidler, *Editors*

Volume 79
Altered Glycosylation in Tumor Cells, Sen-Itiroh Hakomori, Donald M. Marcus, and Christopher L. Reading, *Editors*

Please contact the publisher for information about previous titles in this series.

UCLA Symposia Board

Molecular Biology of Intracellular Protein Sorting and Organelle Assembly

Molecular Biology of Intracellular Protein Sorting and Organelle Assembly

Proceedings of a DuPont-UCLA Symposium Held in Taos, New Mexico
January 30–February 5, 1987

Editors

Ralph A. Bradshaw
Lee McAlister-Henn
Department of Biological Chemistry
University of California, Irvine
Irvine, California

Michael G. Douglas
Department of Biochemistry
University of Texas Health Science Center
Dallas, Texas

Alan R. Liss, Inc. • New York

Address all Inquiries to the Publisher
Alan R. Liss, Inc., 41 East 11th Street, New York, NY 10003

Printed in the United States of America

Library of Congress Cataloging-in-Publication Data

Molecular biology of intracellular protein
sorting and organelle assembly.

(UCLA symposia on molecular and cellular biology ; new ser., v. 76)
"UCLA Symposium on Molecular Biology of Intracellular Protein Sorting and Organelle Assembly"—Contents p.
Includes bibliographies and index.
1. Cytochemistry—Congresses. 2. Cell metabolism—Congresses. 3. Proteins—Congresses. 4. Proteins—Metabolism—Congresses. 5. Cell organelles—Congresses. I. Bradshaw, Ralph A., 1941- . II. McAlister-Henn, Lee. III. Douglas, Michael G. IV. E.I. du Pont de Nemours & Company. V. University of California, Los Angeles. VI. UCLA Symposium on Molecular Biology of Intracellular Protein Sorting and Organelle Assembly (1987 : Taos, N.M.) VII. Series. [DNLM: 1. Cells—metabolism—congresses. 2. Molecular Biology—congresses. 3. Organoids—metabolism—congresses. 4. Proteins—metabolism—congresses.
W3 U17N new ser. v.76 / QU 55 M716 1987]
QH 611.M56 1988 574.87′6042 88-6761
ISBN 0-8451-2675-X

Pages 1–98 of this volume are reprinted from the Journal of Cellular Biochemistry, Volumes 35 and 36. The Journal is the only appropriate literature citation for the articles printed on these pages. The page numbers in the table of contents, contributors list, and index of this volume correspond to the page numbers at the foot of these pages.

The table of contents does not necessarily follow the pattern of the plenary sessions. Instead, it reflects the thrust of the meeting as it evolved from the combination of plenary sessions, poster sessions, and workshops, culminating in the final collection of invited papers, submitted papers, and workshop summaries. The order in which articles appear in this volume does not follow the order of citation in the table of contents. Many of the articles in this volume were published in the Journal of Cellular Biochemistry, and they are reprinted here. These articles appear in the order in which they were accepted for publication and then published in the Journal. They are followed by papers which were submitted solely for publication in the proceedings.

Contents

Contributors

John Armstrong, Department of Biological Sciences, University of Warwick, Coventry CV4 7AL, England; present address: Membrane Molecular Biology Laboratory, Imperial Cancer Research Fund, Lincoln's Inn Fields, London, WC2A 3PX, England **[1]**

Jeffrey Augen, Department of Biochemistry and Molecular Biology, University of Texas Medical School at Houston, Houston, TX 77225 **[109]**

Alison Baker, Department of Biochemistry, Biocenter, University of Basel, CH-4056 Basel, Switzerland **[33]**

Lois M. Banta, Division of Biology, California Institute of Technology, Pasadena, CA 91125 **[173]**

David M. Bedwell, Division of Biology, California Institute of Technology, Pasadena, CA 91125 **[161]**

Phil Bird, Department of Biochemistry, University of Texas Health Science Center at Dallas, Dallas, TX 75235 **[85]**

Ralph A. Bradshaw, Department of Biological Chemistry, College of Medicine, University of California, Irvine, CA 92717 **[xv]**

Marjorie C. Brandriss, Department of Microbiology and Molecular Genetics, University of Medicine and Dentistry of New Jersey-New Jersey Medical School, Newark, NJ 07103 **[105]**

W.R. Brown, Department of Biochemistry and Molecular Biology, The University of Chicago, Chicago, IL 60637 **[57]**

Chuck C.-K. Chao, Department of Biochemistry, University of Texas Health Science Center at Dallas, Dallas, TX 75235; present address: Department of Medicine/ Oncology, Stanford University School of Medicine, Stanford, CA 94305 **[85]**

A.R. Cobitz, Department of Biochemistry and Molecular Biology, The University of Chicago, Chicago, IL 60637 **[57]**

Alan Colman, Department of Biological Sciences, University of Warwick, Coventry CV4 7AL, England **[1]**

Kim E. Creek, National Cancer Institute, Bethesda, MD 20892 **[9]**

Michael G. Cumsky, Department of Molecular Biology and Biochemistry, University of California, Irvine, Irvine, CA **[71]**

R.J. Detrick, Department of Biochemistry and Molecular Biology, The University of Chicago, Chicago, IL 60637 **[57]**

Lori K. Dircks, Department of Molecular, Cellular, and Developmental Biology, University of Colorado, Boulder, CO 80309 **[71]**

Michael G. Douglas, Department of Biochemistry, University of Texas Health Science Center at Dallas, Dallas, TX 75235 **[xv,93]**

Martin Eilers, Department of Biochemistry, Biocenter, University of Basel, CH-4056 Basel, Switzerland **[33]**

Scott D. Emr, Division of Biology, California Institute of Technology, Pasadena, CA 91125 **[161,173]**

Tina Etcheverry, Genentech, South San Francisco, CA 94080 **[153]**

Sharon R. Fried, Department of Biochemistry and Biophysics, University of California, San Francisco, CA 94143 **[187]**

A. Fujiyama, Department of Biochemistry and Molecular Biology, The University of Chicago, Chicago, IL 60637 **[57]**

The number in brackets is the opening page number of the contributor's article.

Reid Gilmore, Department of Biochemistry, University of Massachusetts Medical Center, Worcester, MA 01605 **[123]**

Scott M. Glaser, Department of Molecular Biology and Biochemistry, University of California, Irvine, Irvine, CA **[71]**

L.E. Goodman, Department of Biochemistry and Molecular Biology, The University of Chicago, Chicago, IL 60637 **[57]**

Jeffrey H. Grubb, EA Doisy Department of Biochemistry, St. Louis University School of Medicine, St. Louis, MO 63104 **[9]**

Michael N. Hall, Department of Biochemistry and Biophysics, University of California, San Francisco, CA 94143 **[187]**

William J. Hansen, Department of Biochemistry and Biophysics, University of California Medical School, San Francisco, CA 94143 **[153, 203]**

Franz-Ulrich Hartl, Institut für Physiologische Chemie, Universität München, D-8000 München 2, Federal Republic of Germany **[127]**

Cherie L. Holcomb, Department of Biochemistry, University of California, Berkeley, CA 94720 **[153]**

E.C. Hsueh, Department of Biochemistry and Molecular Biology, The University of Chicago, Chicago, IL 60637 **[57]**

Tsuneo Imanaka, The Rockefeller University, New York, NY 10021 **[139]**

Daniel J. Klionsky, Division of Biology, California Institute of Technology, Pasadena, CA 91125 **[173]**

Paul B. Lazarow, The Rockefeller University, New York, NY 10021 **[139]**

Mark O. Lively, Department of Biochemistry, Bowman Gray School of Medicine, Wake Forest University, Winston-Salem, NC 27103 **[123]**

Debra J. Martin, Department of Biochemistry, University of Iowa College of Medicine, Iowa City, IA 52242 **[99]**

Lee McAlister-Henn, Department of Biological Chemistry, California College of Medicine, University of California, Irvine, CA 92717 **[xv, 105]**

Malcolm McCrae, Department of Biological Sciences, University of Warwick, Coventry CV4 7AL, England **[1]**

Susan Michaelis, Department of Biochemistry and Biophysics, University of California, San Francisco, CA 94143 **[193]**

Walter Neupert, Institut für Physiologische Chemie, Universität München, D-8000 München 2, Federal Republic of Germany **[127]**

Catherine M. Nolan, EA Doisy Department of Biochemistry, St. Louis University School of Medicine, St. Louis, MO 63104 **[9,117]**

Sjur Olsnes, Institute for Cancer Research at The Norwegian Radium Hospital and The Norwegian Cancer Society, Montebello, 0310 Oslo, Norway **[47]**

Ole W. Petersen, Department of Anatomy, The Panum Institute, University of Copenhagen, DK-2200 Copenhagen N, Denmark **[47]**

Nikolaus Pfanner, Institut für Physiologische Chemie, Universität München, D-8000 München 2, Federal Republic of Germany **[127]**

Suzanne R. Pfeffer, Department of Biochemistry, Stanford University School of Medicine, Stanford, CA 94305 **[117]**

Scott Powers, Cold Spring Harbor Laboratory, Cold Spring Harbor, NY 11724 **[193]**

Robert O. Poyton, Department of Molecular, Cellular, and Developmental Biology, University of Colorado, Boulder, CO 80309 **[71]**

Gudepu Radhakrishna, Department of Biochemistry and Molecular Biology, University of Texas Medical School at Houston, Houston, TX 77225 **[109]**

Jane S. Robinson, Division of Biology, California Institute of Technology, Pasadena, CA 91125 **[173]**

Peter A. Rubenstein, Department of Biochemistry, University of Iowa College of Medicine, Iowa City, IA 52242 **[99]**

Kirsten Sandvig, Institute for Cancer Research at The Norwegian Radium Hospital and The Norwegian Cancer Society, Montebello, 0310 Oslo, Norway **[47]**

Randy Schekman, Department of Biochemistry, University of California, Berkeley, CA 94720 **[153]**

Bartholomew M. Sefton, Molecular Biology and Virology Laboratory, Salk Institute, San Diego, CA 92138 **[215]**

Thomas J. Silhavy, Molecular Biology Department, Princeton University, Princeton, NJ 08544 **[133]**

Kai Simons, European Molecular Biology Laboratory, D-6900 Heidelberg, Federal Republic of Germany **[25]**

William S. Sly, EA Doisy Department of Biochemistry, St. Louis University School of Medicine, St. Louis, MO 63104 **[9]**

Cynthia S. Smagula, Department of Biochemistry, University of Texas Health Science Center at Dallas, Dallas, TX 75235 **[93]**

Gillian M. Small, The Rockefeller University, New York, NY 10021 **[139]**

Larry R. Solomon, Department of Biochemistry, University of Iowa College of Medicine, Iowa City, IA 52242 **[99]**

Pamela Stanley, Department of Cell Biology, Albert Einstein College of Medicine, Bronx, NY 10461 **[147]**

Tom H. Stevens, Institute of Molecular Biology, University of Oregon, Eugene, OR 97403 **[117]**

F. Tamanoi, Department of Biochemistry and Molecular Biology, The University of Chicago, Chicago, IL 60637 **[57]**

Cynthia E. Trueblood, Department of Molecular, Cellular, and Developmental Biology, University of Colorado, Boulder, CO 80309 **[71]**

Bo van Deurs, Department of Anatomy, The Panum Institute, University of Copenhagen, DK-2200, Copenhagen N, Denmark **[47]**

Adolphus P.G.M. van Loon, Department of Biochemistry, Biocenter, University of Basel, CH-4056 Basel, Switzerland; present address: Department of Biochemistry, Central Research Unit (ZFE), Hoffmann-La-Roche & Co. AG, CH-4002 Basel, Switzerland **[33]**

Gerrit van Meer, European Molecular Biology Laboratory, D-6900 Heidelberg, Federal Republic of Germany; present address: Department of Cell Biology, Medical Faculty, State University of Utrecht, 3511 HG Utrecht, The Netherlands **[25]**

Keith Verner, Department of Biochemistry, Biocenter, University of Basel, CH-4056 Basel, Switzerland **[33]**

Peter Walter, Department of Biochemistry and Biophysics, University of California, San Francisco, CA 94143 **[203]**

William T. Wickner, Department of Biological Chemistry, UCLA School of Medicine, and Molecular Biology Institute, Los Angeles, CA 90024 **[133]**

Finn Wold, Department of Biochemistry and Molecular Biology, University of Texas Medical School at Houston, Houston, TX 77225 **[109]**

Preface

This volume contains articles presented at the DuPont-UCLA Symposium titled "Molecular Biology of Intracellular Protein Sorting and Organelle Assembly" held at Taos, New Mexico, January 30–February 5, 1987. The last several years have seen major advances in our understanding of how co- and post-translational modifications of proteins, synthesized in the cytoplasm of eukaryotic cells, direct these molecules to their final intra- or extracellular sites of action. The meeting was directed toward examining these events with a view to ascertaining the underlying common principles and mechanisms in these protein trafficking processes. As with all molecular and cell biology, recent advances in technology have had a profound impact on these areas and now allow descriptions in remarkable detail. Control of protein trafficking is among the most fundamental of the molecular events performed by living cells, and therefore these findings have ramifications for a host of disciplines and subdisciplines of biology and human pathology.

The meeting attracted, not unexpectedly, a mix of molecular and cell biologists. Since the boundaries separating these two disciplines have already largely been eroded, there was a great deal of common ground that provided for fruitful discussion and exchange. The more chemically oriented participants described their results in detailed molecular terms, while the cell biologists offered graphic presentations to illustrate how these events fit into the overall metabolism of the cell. The symposium was marked by unusual depth and lively discussion.

This volume is organized largely along the same lines as the meeting. The first half is devoted to co- and post-translational modifications that are directly relevant to protein trafficking and the formation of intracellular organelles. The papers are included under the category of cytoplasmic sorting. The second half of the meeting dealt with intracellular targeting and included events related to the ER/Golgi and recycling through endocytosis. These various presentations were highlighted by several workshops whose summaries are found in appropriate sections. As intended, the composite view is to translate molecular processing into cellular events.

The symposium organizers are particularly grateful to Robin Yeaton and the other members of the UCLA staff for their considerable assistance in the organization and running of the meeting. They also are grateful for the friendly cooperation and warm hospitality offered by the staff of the Kachina Lodge, which hosted this meeting. This was their maiden run for a scientific gathering, and their forbearance and good spirits were fundamental to the success of the event.

The generous Sponsorship of E.I. du Pont de Nemours & Company was integral to the overall success of this meeting. We appreciate additional support provided by Integrated Genetics, Inc., and Genex Corporation.

Ralph A. Bradshaw
Lee McAlister-Henn
Michael G. Douglas

Journal of Cellular Biochemistry 35:129–136 (1987)
Molecular Biology of Intracellular Protein
Sorting and Organelle Assembly 1–8

Expression of Coronavirus E1 and Rotavirus VP10 Membrane Proteins From Synthetic RNA

John Armstrong, Malcolm McCrae, and Alan Colman

Department of Biological Sciences, University of Warwick, Coventry, CV4 7AL, England

Some viruses acquire their envelopes by budding through internal membranes of their host cell. We have expressed the cloned cDNA for glycoproteins from two such viruses, the E1 protein of coronavirus, which buds in the Golgi region, and VP10 protein of rotavirus, which assembles in the endoplasmic reticulum. Messenger RNA was prepared from both cDNAs by using SP6 polymerase and either translated in vitro or injected into cultured CV1 cells or *Xenopus* oocytes. In CV1 cells, the E1 protein was localised to the Golgi region and VP10 protein to the endoplasmic reticulum. In *Xenopus* oocytes, the E1 protein acquired post-translational modifications indistinguishable from the sialylated, O-linked sugars found on viral protein, while the VP10 protein acquired endoglycosidase-H-sensitive N-linked sugars, consistent with their localisation to the Golgi complex and endoplasmic reticulum, respectively. Thus the two proteins provide models with which to study targeting to each of these intracellular compartments. When the RNAs were expressed in matured, meiotic oocytes, the VP10 protein was modified as before, but the E1 protein was processed to a much lesser extent than in interphase oocytes, consistent with a cessation of vesicular transport during cell division.

Key words: expression, Coronavirus E1, Rotavirus VP10, membrane proteins, synthetic RNA

In the eukaryotic cell, the rough endoplasmic reticulum is the site of synthesis for both secretory proteins and integral proteins of the plasma membrane. To reach the cell surface, both types of protein must traverse the Golgi complex. During this transport process, a series of post-translational modifications may occur. Since the modifying enzymes are localised along the secretory pathway, the endoplasmic reticulum and Golgi complex constitute a series of membrane-limited compartments, each apparently with a distinct complement of proteins [reviewed in 1–4]. How are these proteins confined to their appropriate destinations, rather than migrating onward to the plasma membrane? We are investigating two viral model proteins, one for the

John Armstrong's present address is Membrane Molecular Biology Laboratory, Imperial Cancer Research Fund, PO Box 123, Lincoln's Inn Fields, London, WC2A 3PX, UK.

Received March 20, 1987; accepted May 20, 1987.

endoplasmic reticulum and one for the Golgi complex, with a view to determining the features of each molecule responsible for its correct localisation.

Rotaviruses are a class of animal viruses which characteristically assemble at the endoplasmic reticulum [5]. Although assembly appears to be a budding process, the matured virion surprisingly lacks a visible envelope [5–7]. The bovine UK strain encodes two glycoproteins: VP7, which forms part of the virion, and VP10, which does not [8,9]. The latter is an integral membrane glycoprotein [9,10] bearing "immature" N-linked sugars which imply a failure to leave the endoplasmic reticulum [10].

By contrast, with coronaviruses the envelope is acquired by budding initially at smooth membranes close to the Golgi complex but continuous with the endoplasmic reticulum; as infection progresses budding may also occur in the Golgi cisternae or the rough endoplasmic reticulum [11,12]. The strain MHV-A59 bears two glycoproteins, the smaller of which, E1, is restricted within the cell and does not appear to reach the plasma membrane except as part of a budded virion [12,13].

Thus we have adopted the proteins VP10 and E1 as potential models for the class of membrane proteins which localise within compartments of the secretory pathway. Previously we have reported the cloning and sequence analysis of both proteins [9,14]. Here we show that both cDNAs can be expressed via artificial mRNA prepared using SP6 polymerase [15,16]. The mRNA may be translated either in vitro or after injection into *Xenopus* oocytes or cultured CV1 cells. Expressed in this way, VP10 and E1 prove to be valid model proteins for the endoplasmic reticulum and Golgi complex, respectively. In addition, we have exploited the fact that *Xenopus* oocytes can be matured into a meiotic state [17,18] to investigate the transport of the two proteins to their destinations during cell division.

MATERIALS AND METHODS

Construction of Recombinant Plasmids and Synthesis of RNA

All restrictions, ligations, and transformations were carried out according to standard methods [19]. Full-length cDNA for bovine rotavirus VP10 (assembled and kindly provided by H. Baybutt) was excised with the enzymes AhaIII and SalI, to give a fragment corresponding to nucleotides 8-588 [in Figure 3 of 9]. This was inserted into the BglII site of the transcription vector pSP64T [15].

E1 cDNA [14,20] was excised with AhaIII and FokI, representing nucleotides 59–780 [in 14], plus the following 13 bases of the adjacent nucleocapsid gene [21] and inserted into the BglII site of pSP64T [15].

Both plasmids were linearised with EcoRI and RNA transcribed with SP6 polymerase in the presence of 0.5mM m^7G (5′)ppp(5′)G exactly as described [16]. Since the vector pSP64T contains a poly-A tract between the BglII and EcoRI sites, this method results in the synthesis of capped, polyadenylated mRNA in a single reaction.

Translation of RNA and Analysis of Products

Translation in vitro in reticulocyte lysates, injection of *Xenopus* oocytes, immunoprecipitation of proteins, microinjection of CV1 cells, and immunofluorescence analysis were all performed as before [16]. For analysis of VP10 protein, rabbit antiserum to a fusion protein of bacterial β-galactosidase and VP10, whose preparation will be described in detail elsewhere, was used. To detect E1, we used a rabbit serum to E1 purified by detergent extraction of virus [22], kindly provided by S. Tooze, EMBL, Heidelberg, who also provided stocks of Coronavirus MHV-A59 and its host, sac$^-$ cells. Radiolabelled virus was produced by infection of cells at a ratio of 10 pfu/cell, incubation for 16 hr in methionine-free medium to which was added 50 μCi/ml ^{35}S-methionine (Amersham), and collection of the culture supernatant. Endoglycosidase H digestions were carried out as before [18].

Xenopus oocytes were matured into second meiotic metaphase by incubation in 10 μg/ml progesterone as described [18].

RESULTS

Expression of VP10 and E1 From Synthetic RNA

Previously we showed that RNA prepared with SP6 polymerase could be translated in cultured cells, provided it has both a 5′ cap structure, and a 3′ poly-A tract added using poly-A polymerase [16]. E1 RNA prepared in this way was found to be translated in both CV1 cells and *Xenopus* oocytes (not shown). We have further simplified this method by using a vector which includes a sequence of 23 A's in the transcribed region [15], eliminating the requirement for a second reaction. RNAs prepared by this method for VP10 and E1 were translated efficiency in reticulocyte lysates, *Xenopus* oocytes, and cultured CV1 cells (Figs. 1, 2).

Analysis of VP10 and E1 From Oocytes

VP10 protein produced during viral infection bears N-linked oligosaccharides of the form $Man_8GlcNac_2$ [10], characteristic of proteins which have not left the

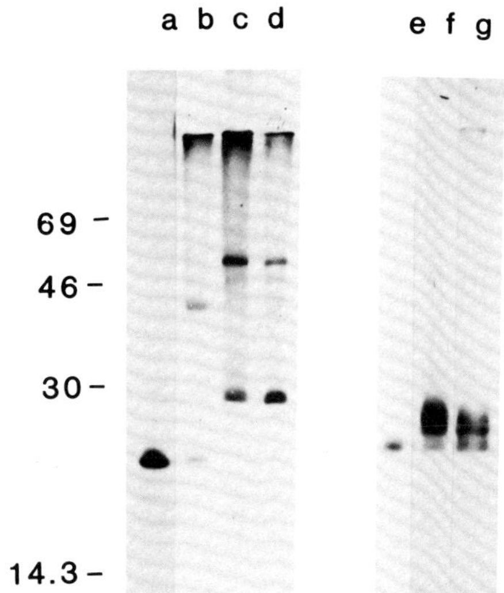

Fig. 1. Expression of VP10 and E1 in *Xenopus* oocytes. **a:** VP10 protein translated in a reticulocyte lysate. **b–d:** VP10 protein immunoprecipitated from injected oocytes and digested with endoglycosidase H (b), mock-digested (c), or untreated (d). The identity of the high molecular weight band is unknown. **e:** E1 protein translated in a reticulocyte lysate. **f:** E1 protein immunoprecipitated from injected oocytes. **g:** E1 protein immunoprecipitated from radiolabelled viral supernatant. Proteins were separated on a 15% acrylamide SDS gel and fluorographed for 18 hr (b–e) or 1 wk (a,f,g). Numbers on the left are molecular weights in kilodaltons of marker proteins.

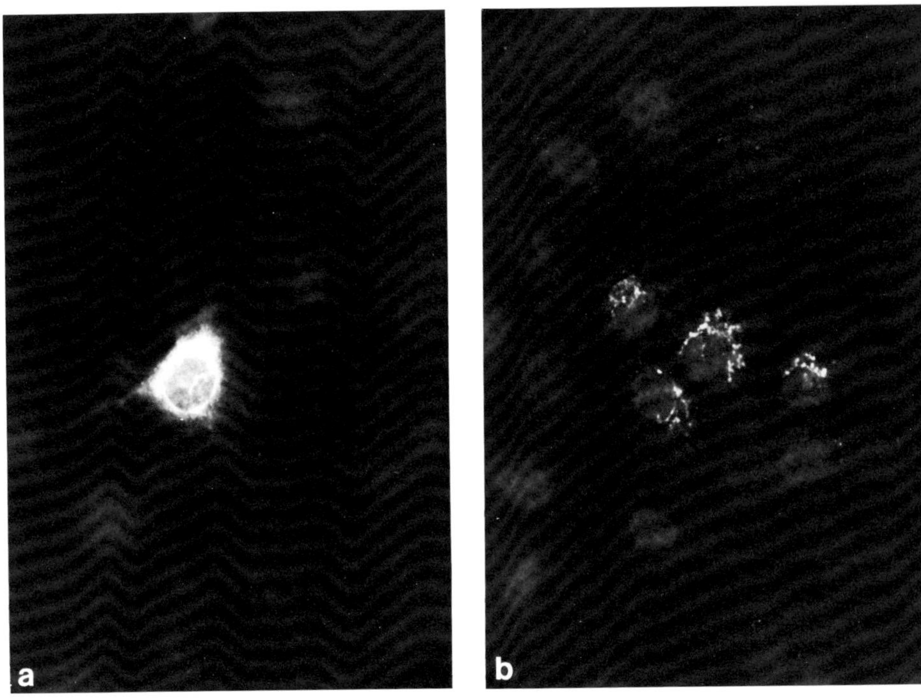

Fig. 2. Expression of VP10 and E1 in CV1 cells. VP10 (**a**) or E1 (**b**) RNAs were injected into cells which were incubated for 8 hr, fixed, permeabilised with Triton X-100, and treated with anti-VP10 (1/100) or anti-E1 (1/40) followed by fluorescein-conjugated antirabbit IgG (1/40). Cells were photographed under epifluorescence using a Nikon Optiphot microscope.

rough endoplasmic reticulum; these structures may be cleaved by endoglycosidase H [23]. VP10 protein which had been immunoprecipitated from injected oxcytes was found to be completely sensitive to endoglycosidase H, yielding a species which approximately comigrated with the unprocessed form produced in the reticulocyte lysate (Fig. 1, lanes a–d). Thus the VP10 protein is likely to be localised within the endoplasmic reticulum of the oocyte.

E1 protein was synthesized in oocytes as a spectrum of forms, the most mobile of which comigrated with the unmodified protein from the reticulocyte lysate (Fig. 1, lanes e,f). The oocyte proteins were very similar in both mobilities and relative abundance to the species of E1 found in virus particles (Fig. 1, lane g). These have been shown to contain O-linked sugars as their only posttranslational modification [24,25]. Since at least the later stages of O-linked glycosylation are thought to occur in the Golgi complex [26], most or all of the E1 protein would appear to have reached this organelle in the oocyte.

With both antibodies, no significant labeled proteins were precipitated from oocytes in the absence of the appropriate mRNA.

Analysis of VP10 and E1 in CV1 Cells

SP6 RNAs for VP10 and E1 were microinjected into CV1 monkey kidney cells, and the resulting proteins were detected by immunofluorescence. VP10 proteins was found in an elaborate pattern around the nucleus and throughout the cytoplasm, characteristic of the rough endoplasmic reticulum (Fig. 2a). By changing the plane of focus, it became clear that all of the nuclear envelope was labeled (not shown). This was expected as the outer nuclear membrane is continuous with, and usually considered as part of, the endoplasmic reticulum. In contrast, E1 protein showed a much more localised fluorescence pattern (Fig. 2b). Labeling was concentrated in a perinuclear area corresponding to the Golgi region and coincident with the intracellular pattern shown by fluorescent wheat germ agluttinin, a Golgi marker (not shown). The pattern also closely resembled the distribution of E1 protein in virally infected cells [12,13]. If injected cells were not permeabilised with detergent, no labeling of the cell surface was detected with either protein (not shown).

Expression of VP10 and E1 in Meiotic Oocytes

Incubation of *Xenopus* oocytes in progesterone causes them to mature from their initial stage, first meiotic prophase, to second meiotic metaphase [17]. In this state, the Golgi apparatus is broken down and protein secretion blocked [18]. We investigated the synthesis of VP10 and E1 in such matured oocytes. VP10 was electrophoretically indistinguishable from the form produced in nonmatured oocytes (Fig. 3, lanes a,b). E1, in contrast, showed a striking reduction in the extent to which it was processed in comparison to nonmatured oocytes (Fig. 3, lanes c,d), implying that most of the protein is denied access to the modifying enzymes of the Golgi complex under these conditions.

DISCUSSION

Transcription of RNA with SP6 polymerase has been used as a route for expression of cloned cDNAs either in vitro or in the *Xenopus* oocyte [15]. If the RNA incorporates a 5′ cap structure and a 3′ poly-A tract, it may also be expressed in

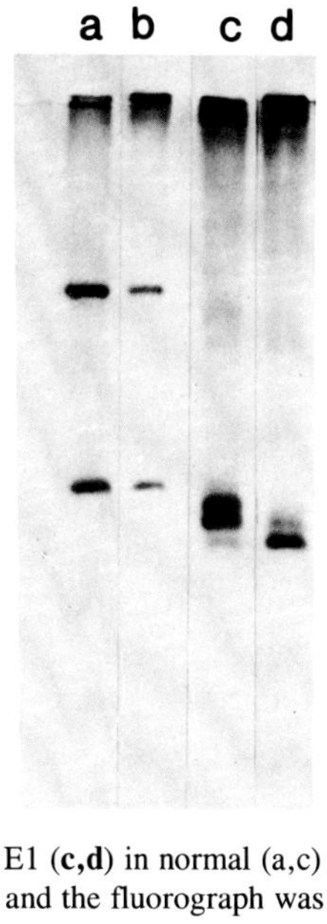

Fig. 3. Expression of VP10 (**a,b**) and E1 (**c,d**) in normal (a,c) or progesterone-matured (b,d) oocytes. Proteins were analysed as for Figure 1, and the fluorograph was exposed for 1 wk.

cultured cells [16]. By using the vector pSP64T [15], such RNA can be prepared in a single reaction. We assume that the presence of an encoded poly-A sequence in the vector is the critical factor in allowing expression of the RNA in cultured cells [16]; however, the transcribed region also contains 5′ and 3′ untranslated regions from a globin cDNA, and a sequence of C bases at the extreme 3′ end [15], and any of these elements may contribute to the stability of the mRNA or the efficiency with which it is translated. Whichever feature is important, the system provides a quick and versatile approach to the expression of cloned DNAs. Its principal limitation, at present, is the lack of a technique for the efficient introduction of RNA into large populations of cultured cells.

It appears that both of the proteins we have studied, VP10 and E1, are localised in a similar fashion whether they are expressed in isolation in cell types as diverse as oocytes and CV1 cells, or in the course of viral infection. Thus each is likely to be a legitimate model with which to study targeting of proteins to its respective intracellular destination. In the case of VP10, this is the rough endoplasmic reticulum, as judged by fluorescence microscopy and sensitivity of its N-linked oligosaccharides to endoglycosidase H (Figs. 1, 2). The corresponding protein of rotavirus SA11, termed NCVP5, bears oligosaccharides of the form $Man_8GlcNAc_2$ [10].

In contrast, the principal location of the E1 protein appears to be the Golgi complex. As the bulk of the protein made in oocytes has been modified to give forms of similar mobilities to the viral protein (Fig. 1, lanes f,g; see also [27]), it has presumably acquired O-linked oligosaccharides, an indication of having reached the Golgi complex [26]. Most of the oligosaccharides on viral E1 contain terminal sialic acid [25], whose presence is detectable by changes in electrophoretic mobility of the glycoprotein after neuraminidase digestion. However, we have repeatedly failed to observe any effect on mobility of oocyte E1 by treatment with neuroaminidases from various sources and under various conditions (not shown). The explanation for this is not clear; perhaps the oocyte adds exotic sialic acids which are not susceptible to cleavage.

Immunofluorescence of cultured cells expressing E1 clearly showed a very localised perinuclear distribution (Fig. 2b) characteristic of the Golgi region and similar to the pattern observed during viral infection [12]. However, not all of the viral protein is restricted to the flattened cisternae of the Golgi complex; some at least is found in smooth membranes which are in the same region of the cell but are in fact continuous with the rough endoplasmic reticulum [12,22]. These membranes are reminiscent of the transitional elements of specialised secretory cells [28] and are the earliest site of budding during viral infection [12]. At the resolution of light microscopy it is impossible to say whether the pattern in Figure 2b includes this compartment. It will be of interest to resolve this point by immunoelectron microscopy, which should also allow us to determine how many of the Golgi cisternae are labelled. Nevertheless, it is probably safe to conclude that the behaviour of the E1 protein is largely responsible for determining the intracellular budding site of coronavirus.

The localisation of VP10 might argue for a similar function in rotavirus morphogenesis. However, this protein's precise role in viral infection remains enigmatic. The other rotavirus glycoprotein, VP7, also appears to be restricted to the endoplasmic reticulum when it is expressed in the absence of other viral proteins [29]. Thus both proteins may be involved in determining the viral assembly site, but the subsequent fate of VP10 is unclear, since it is not detectable in purified virions [9].

Perhaps the simplest hypothesis is that the newly budded virus has a lipid envelope including VP10 and VP7, which is somehow shed at a later stage to leave VP7 connected directly to components of the capsid [7].

An aspect of the biogenesis of the endoplasmic reticulum and Golgi complex which has recently attracted attention is their fate during cell division. In mammalian cells, both organelles are thought to break up into vesicles which then partition randomly between the daughter cells and fuse together; concomitant with this disruption is a cessation of traffic of secretory and plasma mebmrane proteins to and through the Golgi complex [reviewed in 30]. The same phenomena appear to occur in the *Xenopus* oocyte, which has the experimental advantage that its cell cycle state may be manipulated with hormones [18]. The results presented in Figure 3 are entirely consistent with this model; VP10 is synthesized and processed normally in meiotic oocytes, but E1 is processed to a far lesser extent than in interphase oocytes, and thus has probably failed to traverse what is assumed to be the first vesicle-mediated step of membrane traffic, from the endoplasmic reticulum to the Golgi complex. It might be of interest to study the vesicles in which E1 becomes trapped under these conditions.

What features of each protein might be involved in determining its intracellular destination, and what are the mechanisms involved in each case? The two proteins share a superficial topological similarity, in having a very short domain on the lumenal side of the membrane which is N-terminal but does not arise from cleavage of a signal sequence; E1 has the additional unusual feature of hydrophobic regions large enough to span the membrane three times [9,10,14,31,32].

Two mechanisms have been proposed for retention of proteins in the endoplasmic reticulum. One is that a species variously known as BiP or GRP78 attaches to improperly folded or assembled molecules and prevents their exit to the Golgi complex [33–35]. The second is that all of the membrane proteins of the endoplasmic reticulum form a continuous interacting network and are unable to diffuse in the lipid bilayer [36,37]. In respect of either model, we have never observed any host species which coprecipitates from oocytes with VP10, although admittedly our labelling schedule is perhaps not optimal for detection. (Note that the unidentified band in Figure 1, lanes b–d is made only in response to VP10 mRNA.) Neither have we observed such species with our Golgi model protein, E1, whose sorting mechanism is even more obscure. However, with the availability of cloned cDNA and a reliable expression system, it is now possible to dissect each molecule and identify the characteristics involved in its correct targeting within the cell.

ACKNOWLEDGMENTS

This work was supported by grants from the Cancer Research Campaign (to A. Colman) and the Wellcome Trust (to A. Colman and M. McCrae).

REFERENCES

1. Kelly RB: Science 230:25, 1985.
2. Rothman JE, Lenard J: Trends Biochem Sci 9:176, 1984.
3. Kornfeld R, Kornfeld S: Annu Rev Biochem 54:631, 1985.
4. Roman LM, Garoff H: Trends Biochem Sci 10:428, 1985.
5. Altenburg BC, Graham DY, Estes MK: J Gen Virol 46:75, 1980.
6. Soler C, Musalem C, Lorono M, Espejo RT: J Virol 44:983, 1982.

7. Petrie BL, Estes MK, Graham DY: J Virol 46:270, 1983.
8. McCrae MA, Faulkner-Valle GP: J Virol 39:490, 1981.
9. Baybutt HN, McCrae MA: Virus Res 1:533, 1984.
10. Kabcenell AK, Atkinson PH: J Cell Biol 101:1270, 1985.
11. David-Ferreira JF, Manaker R: J Cell Biol 24:57, 1965.
12. Tooze J, Tooze S, Warren G: Eur J Cell Biol 33:281, 1984.
13. Dubois-Dalcq ME, Doller EW, Haspel MV, Holmes KV: Virology 119:317, 1982.
14. Armstrong J, Niemann H, Smeekens S, Rottier P, Warren G: Nature 308:751, 1984.
15. Krieg PA, Melton DA: Nucleic Acids Res 12:7057, 1984.
16. Drummond DR, Armstrong J, Colman A: Nucleic Acids Res 13:7375, 1985.
17. Smith LD: In Weber R (ed): "Biochemistry of Animal Development," Vol. 3, New York: Academic Press, 1975, pp 1–46.
18. Colman A, Jones EA, Heasman J: J Cell Biol 101:313, 1985.
19. Maniatis T, Fritsch EF, Sambrook J (eds): "Molecular Cloning: A Laboratory Manual." Cold Spring Harbor, NY: Cold Spring Harbor Laboratory, 1982.
20. Niemann H, Heisterberg-Moutsis G, Geyer R, Klenk H-D, Wirth M: In Rottier PJM, Van der Zeijst BAM, Spaan WJM, Horzinek MC (eds): "Molecular Biology and Pathogenesis of Coronaviruses." New York: Plenum Press, 1984, pp 201–213.
21. Armstrong J, Smeekens S, Rottier P: Nucleic Acids Res 11:883, 1983.
22. Tooze S, Stanley KK: J Virol 60:928, 1986.
23. Tarentino AL, Maley F: J Biol Chem 249:811, 1974.
24. Niemann H, Klenk H-D: J Mol Biol 153:993, 1981.
25. Niemann H, Geyer R, Klenk H-D, Linder D, Stirm S, Wirth M: EMBO J 3:665, 1984.
26. Niemann H, Boschek B, Evans D, Rosing M, Tamura T, Klenk H-D: EMBO J 1:1499, 1982.
27. Rottier PJM, Spaan WJM, Horzinek MC, Van der Zeijst BAM: J Virol 38:20, 1981.
28. Palade G: Science 189:347, 1975.
29. Poruchynsky MS, Tyndall C, Both GW, Sato F, Bellamy AR, Atkinson PH: J Cell Biol 101:2199, 1985.
30. Warren G: Trends Biochem Sci 10:439, 1985.
31. Ericson BL, Graham DY, Mason BB, Hanssen HH, Estes MK: Virology 127:320, 1983.
32. Rottier P, Brandenburg D, Armstrong J, Van der Zeijst B, Warren G: Proc Natl Acad Sci USA 81:1421, 1984.
33. Bole DG, Hendershot LM, Kearney JF: J Cell Biol 102:1558, 1986.
34. Munro S, Pelham HRB: Cell 46:291, 1986.
35. Gething M-J, McCammon K, Sambrook J: Cell 46:939, 1986.
36. Hortsch M, Avossa D, Meyer DI: J Cell Biol 103:241, 1986.
37. Kreibich G, Ulrich BL, Sabatini DD: J Cell Biol 77:464, 1978.

Journal of Cellular Biochemistry 35:137–151 (1987)
Molecular Biology of Intracellular Protein
Sorting and Organelle Assembly 9–23

Antibody to the Phosphomannosyl Receptor Inhibits Recycling of Receptor in Fibroblasts

Catherine M. Nolan, Kim E. Creek, Jeffrey H. Grubb, and William S. Sly

EA Doisy Department of Biochemistry, St Louis University School of Medicine, St Louis, Missouri 63104 (C.M.N., J.H.G., W.S.S.), National Cancer Institute, Bethesda, Maryland 20892 (K.E.C.)

The 215-kd phosphomannosyl receptor is involved in the transport of newly synthesized acid hydrolases to lysosomes and also mediates the pinocytosis of lysosomal enzymes by fibroblasts in culture. Recycling of receptors to the sorting sites is an integral part of both these processes. In this report, we describe the inhibition in human fibroblasts of both functions of the phosphomannosyl receptor by a rabbit antiserum to the bovine liver receptor. This inhibition cannot be completely accounted for by inhibition of ligand-receptor interaction. Rather the antibody appears to cross-link receptors and cause a removal of receptors from the sorting sites (plasma membrane and Golgi apparatus) and their accumulation in a compartment from which they do not recycle. Removal of receptors from the recycling pool by antibody is irreversible, and return of receptors requires synthesis of new protein. Degradation of "trapped receptors" is enhanced (t½ = 7.5 hr), but much more gradual than their removal from the functional receptor pool (t½ = 30 min).

Key words: receptor inactivation, lysosomal enzyme targeting

The 215-kd phosphomannosyl receptor (PMR), or cation-independent phosphomannosyl receptor, is an integral membrane glycoprotein known to be involved in the targeting of newly synthesized acid hydrolases to lysosomes and in endocytosis of lysosomal enzymes by fibroblasts in culture [1–3]. The receptor has been identified in and purified from many sources including bovine liver [4], Swarm rat chondrosarcoma membranes [5], rat liver [6], and human liver [7]. Antibodies to the receptor have been raised in several laboratories [6–9].

We have used a rabbit antibody to the bovine liver PMR to study the biosynthesis and turnover of the PMR in human fibroblasts and have shown that the receptor has a half-life of 24–29 hr [8]. The receptor is also turned over slowly in monolayers of Chinese hamster ovary cells [10]. The slow turnover of the receptor, together with

Received January 27, 1987; revised and accepted May 11, 1987.

previous kinetic data of ligand internalization, led to the conclusion that, following receptor-mediated endocytosis, the ligand is delivered to lysosomes and the receptor recycles to the plasma membrane to be reutilized [11,12]. Inhibition of recycling by lysosomotropic amines led to the suggestion that pH-dependent release of ligand in an acidic compartment is required for receptors to recycle [12]. The effects of amines on the transport of newly synthesized enzymes suggested that receptor recycling is also an essential feature of transport of newly synthesized acid hydrolases to lysosomes [12,13].

We have previously observed that antibodies to the bovine liver PMR inhibited PMR-mediated endocytosis of β-glucuronidase by human fibroblasts [8]. In this report we further characterize this inhibition and describe how these antibodies also inactivate the targeting of newly synthesized acid hydrolases. We present evidence to suggest that the inactivation of PMR function can be explained by the failure of receptors cross-linked by antibody to recycle to the sites of enzyme sorting (plasma membrane and Golgi apparatus).

MATERIALS AND METHODS

Materials

Goat antirabbit IgG (heavy- and light-chain specific) and fluorescein-conjugated $F(ab^1)_2$ fragment rabbit antigoat IgG (heavy- and light-chain specific) were from Cappel Laboratories (Melvern, PA). [2-^{3}H]-Mannose, (15.8 Ci/mm), [^{35}S]-methionine (953 Ci/mmol), carrier-free Na^{125}I (17 Ci/mg), and EN^{3}HANCE were from New England Nuclear (Boston, MA). β-Nicotinamide adenine dinucleotide, reduced form, and epidermal growth factor (EGF) from mouse submaxillary glands, were from Sigma Chemical Co. (St Louis, MO). Fluorometric substrates were obtained from Research Products International (Elk Grove, IL). Mercuripapain was from Worthington (Freehold, NJ). Iodo-Beads were from Pierce Chemical Co. (Rockford, IL). β-Glucuronidase was purified to homogeneity from human spleen as described previously [14] and an affinity-purified, high-uptake fraction was obtained by chromatography on immobilized phosphomannosyl receptors [15].

Cells and Cell Culture

Normal diploid human fibroblasts (IMR-91), β-glucuronidase-deficient fibroblasts (GM-2784), and Tay-Sachs disease fibroblasts (GM-502) were received from the Human Mutant Cell Repository (Camden, NJ). β-Hexosaminidase-deficient fibroblasts (WH 98/EW) were received from the Repository for Mutant Human Cell Strains (Montreal, Quebec, Canada). Cultures were maintained at 37°C in 5% CO_2 in minimal essential medium MEM, (Earle's Salts, Gibco) containing 15% fetal calf serum, 1 mM sodium pyruvate, 100 units/ml penicillin, and 100 μg/ml streptomycin sulfate.

Preparation of Antibodies

Antibodies to purified bovine liver phosphomannosyl receptor were raised in a rabbit as previously described [8]. An IgG fraction (anti-PMR IgG) was isolated from immune serum by two methods: 1) ammonium sulfate precipitation and chromatography on DEAE-52 [16], and 2) chromatography on protein-A Sepharose CL-4B (Pharmacia, Piscataway, NJ) using 3 M KSCN as eluant, as recommended by the manufacturer.

Anti-PMR Fab fragments were prepared from IgG obtained by the latter method, by papain-digestion [17]. Fab fragments were separated from Fc fragments and from intact IgG molecules by chromatography on protein A Sepharose. Lack of contamination of Fab fragments by IgG was ascertained by sodium dodecyl sulfate-polyacrylamide gel electrophoresis (SDS-PAGE) on a 10% gel [18].

Goat antihuman PMR antibodies were raised by injecting a goat with PMR purified from human liver by the method of Sahagian [9]. Homogeneity of the receptor preparation was confirmed by SDS-PAGE [18]. The antiserum was shown to specifically recognize PMR in human fibroblasts by immunoprecipitation of the receptor from metabolically labeled cells and by immunoblotting of fibroblast membranes (data not shown).

Rabbit antiserum to purified human placental β-hexosaminidase was prepared by Dr. Robert Nicolotti and an IgG fraction was prepared from the antiserum by ammonium sulfate precipitation and chromatography on DEAE Affi-Gel Blue (Bio-Rad Laboratories, Richmond, CA).

Enzyme and Protein Assays

β-Glucuronidase and β-hexosaminidase activities were determined as described previously [19,20]. Lactate dehydrogenase (LDH) activity was measured spectrophotometrically, as described by Reeves et al [21]. Protein was measured by the method of Lowry et al [22] using bovine serum albumin as standard.

Ligand Iodination

EGF and affinity-purified β-glucuronidase were iodinated with Iodo-Beads, according to the manufacturer's recommendations, to specific radioactivities of 47×10^6 cpm/μg and 39×10^6 cpm/μg, respectively.

Ligand Pinocytosis by Fibroblasts

Pinocytosis of β-glucuronidase was determined in duplicate 35-mm petri dishes containing β-glucuronidase-deficient fibroblasts at confluence (0.2–0.3 mg protein/dish) as previously described [12,23]. Pinocytosis of ^{125}I-EGF was determined in similar dishes by incubating cells with medium containing 10 ng/ml of ^{125}I-EGF, for 30 min at 37°C. Cells were subsequently chilled on ice, rinsed five times with PBS, solubilized in 0.5 ml of 1% sodium deoxycholate, and counted for ^{125}I.

Uptake of β-hexosaminidase was conducted in β-hexosaminidase-deficient Sandhoff disease fibroblasts grown to confluence in 35-mm petri dishes. β-Hexosaminidase (500 units) was added to the cells in 1.0 ml of Eagle's minimal essential medium and incubation was for 2 hr at 37°C. Cell-associated β-hexosaminidase was measured in cells disrupted by freeze-thaw in 0.5 ml of H_2O.

Binding of ^{125}I-β-Glucuronidase

Preparation of total cell membranes and determination of ^{125}I-β-glucuronidase binding to these were as described previously [19]. For determination of cell-surface binding of ^{125}I-β-glucuronidase, confluent monolayers of fibroblasts in 35-mm petri dishes were chilled to 4°C and incubated with 1 ml medium (MEM, 5 mg/ml human serum albumin [HSA], ± 10 mM mannose 6-phosphate (M6-P) containing 250,000 cpm ^{125}I-β-glucuronidase/4,000 units enzyme activity, for 1 hr at 4°C. The monolayers were then washed five times with cold PBS, solubilized in 1 ml 1% DOC, and counted for ^{125}I. Specific binding was estimated from binding in the presence and absence of M6-P.

Radiolabeling and Immunoprecipitation of β-Hexosaminidase B From Tay-Sachs Disease Fibroblasts

Tay-Sachs disease fibroblasts were grown to confluence in 100-mm petri dishes (Corning). Prior to labeling the cultures were rinsed once with 5.0 ml of low glucose (.28 mM) MEM containing HSA (1 mg/ml), glutamine (2 mM), and pyruvate (0.1 mg/ml) and starved for 30 min in 5.0 ml of this medium at 37°C in a 5% CO_2 atmosphere. The fibroblasts were then labeled for 21 hr in 5.0 ml of the above medium containing [2-^{3}H]mannose, (60 μCi/ml), mannose 6-phosphate (2 mM), and either NH_4Cl (10 mM) or anti-PMR IgG (150 μg/ml).

At the end of the labeling period, the medium was removed and precipitated with 2.5 g of solid $(NH_4)_2SO_4$ for 1 hr at 4°C. The precipitate was collected by centrifugation at 10,000g for 10 min and the pellet was resuspended in 0.5 ml of Dulbecco's phosphate-buffered saline (PBS). Labeled β-hexosaminidase in the medium extract was immunoprecipitated as described below.

Following radiolabeling the cell monolayers were washed twice with 5.0 ml of PBS; the cells were scraped from the dish in PBS with a rubber policeman and pelleted by low-speed centrifugation. The cell pellets were carefully resuspended in PBS and washed a second time. After washing, the cells were resuspended in 0.5 ml of 50 mM Tris, pH 7.0, 10 mM mannose 6-phosphate, 5 mM EDTA, and 10 μg/ml aprotinin and disrupted by three 10-sec bursts with a sonicator (Bronwill Scientific). Sodium deoxycholate was then added to a final concentration of 0.025%; the disrupted cells were placed at 4°C for 30 min, and the membranes were removed by centrifugation at 30,000g for 60 min. The supernatant was used for immunoprecipitation of cell-associated β-hexosaminidase B. Under these conditions greater than 90% of the total cellular β-hexosaminidase is recovered in the cell extract and greater than 90% of the secreted β-hexosaminidase is recovered in the $(NH_4)_2SO_4$ precipitate.

Immunoprecipitation of radiolabeled cellular and secreted β-hexosaminidase B was conducted as follows. [2-^{3}H]Mannose-labeled cell extracts or medium extracts (450 μl) were added to 45 μl of 10% Nonidet P-40, 10% sodium deoxycholate, 1% sodium dodecyl sulfate (SDS) in PBs. Formalin-fixed *Staphylococcus aureus* (50 μl of a 10% v/v suspension) was added and pelleted after 30 min at room temperature. The supernatants were transferred to new tubes and antigen-antibody complexes were formed by the addition of 25 μl of rabbit antihuman placental β-hexosaminidase antibodies (45 μg of IgG). After 2 hr at room temperature, 50 μl of formalin-fixed *S. aureus* was added, and immune complexes were collected 30 min later by centrifugation for 1 min in a Beckman microfuge. These complexes were resuspended in 1.0 ml of 0.5% Nonidet P-40, 0.5% sodium deoxycholate, and 0.05% SDS in PBS, transferred to a new tube, and washed two more times with 0.5 ml of this detergent-containing buffer, once with this buffer containing 350 mM added NaCl, and finally once with 0.5 ml of PBS. The immunoprecipitates were released from the *S aureus* by boiling for 5 min in 100 μl of sample buffer (75 mM Tris; pH 6.8, 2% SDS, 5% 2-mercaptoethanol, 5% glycerol, 0.0015% bromophenol blue). Polyacrylamide gel electrophoresis was conducted under reducing conditions in 10% polyacrylamide gels containing 0.1% (w/v) SDS as described by Laemmli [18]. Bio-Rad (Richmond, CA) low molecular weight standards were visualized by staining with Coomassie Blue.

Gels were subsequently processed for fluorography in EN^3HANCE and exposed to Kodak XSAR-5 film at −70°C for 9 days.

Immunofluorescence

Subconfluent fibroblasts on tissue culture chamber slides were fixed and permeabilized in 100% methanol at 4°C for 10 min. They were incubated for 1 hr, at room temperature, with goat antihuman liver PMR serum (diluted 1:30 in PBS containing 1% normal rabbit serum and 0.2% gelatin) followed by fluorescein-conjugated rabbit F $(ab^1)_2$ antigoat IgG (diluted 1:20 in the same buffer) and examined and photographed with a Leitz orthoplan fluorescent microscope with a filter suitable for green fluorescence (520 nm).

In control experiments, it was shown that binding of rabbit antibovine liver PMR by fibroblasts did not interfere with subsequent detection of PMR in these cells by the goat anti-PMR serum.

RESULTS

Figure 1A shows the inhibition of β-glucuronidase uptake with increasing concentrations of antibovine liver-PMR serum in the medium. At an antiserum concentration of 40 μl/ml, β-glucuronidase uptake is less than 20% that of control, non-antibody-treated cells. Preimmune serum showed no inhibitory effect on β-glucuronidase uptake (data not shown). Figure 1A also shows that the uptake of ^{125}I-EGF, another molecule internalized by receptor-mediated endocytosis, is unaffected. Thus, the inhibitory effect on pinocytosis is specific for PMR-mediated pinocytosis.

In order to determine whether inhibition of enzyme pinocytosis was secondary to blocking of the ligand binding site of the PMR by the antibody, we measured the effect of anti-PMR antiserum on 4°C cell-surface binding, using ^{125}-I-β-glucuronidase. As shown in Figure 1B, anti-PMR serum interfered with binding to a maximum of 30%. This interference, however, was barely, if at all, greater than that seen with preimmune serum and could not fully account for the observed inhibition of endocytosis. The inhibition of binding by preimmune serum may be due to competition from other mannose-6-P (M6-P)-containing acid hydrolases which are released from platelets during the clotting process in the preparation of rabbit serum.

The almost complete inhibition of ligand internalization, in the absence of a comparably large effect on ligand binding, suggested the possibility of antibody-induced removal of receptors from the cell surface. To test this, we incubated cells at 37°C with a 50 μl/ml concentration of anti-PMR serum for various time intervals and measured the cell-surface enzyme binding activity and enzyme pinocytosis (Fig. 2). Both activities decreased in a similar, time-dependent manner, with the half-life of both processes being 25–30 min. This is consistent with loss of pinocytosis activity being a consequence of disappearance of receptors from the cell surface.

Several studies with other ligands and their receptors have shown a decrease in cell-surface receptor number on treatment of cells with antireceptor antibodies [24–30]. In some cases, this decrease has been ascribed to antibody-induced cross-linking of receptors and their subsequent failure to recycle to the surface [24–27]. To see if antibody-induced cross-linking of PMR might be a factor in the inactivation of pinocytosis of β-glucuronidase, we prepared anti-PMR Fab fragments by papain digestion of anti-PMR IgG and examined their effects on β-glucuronidase uptake.

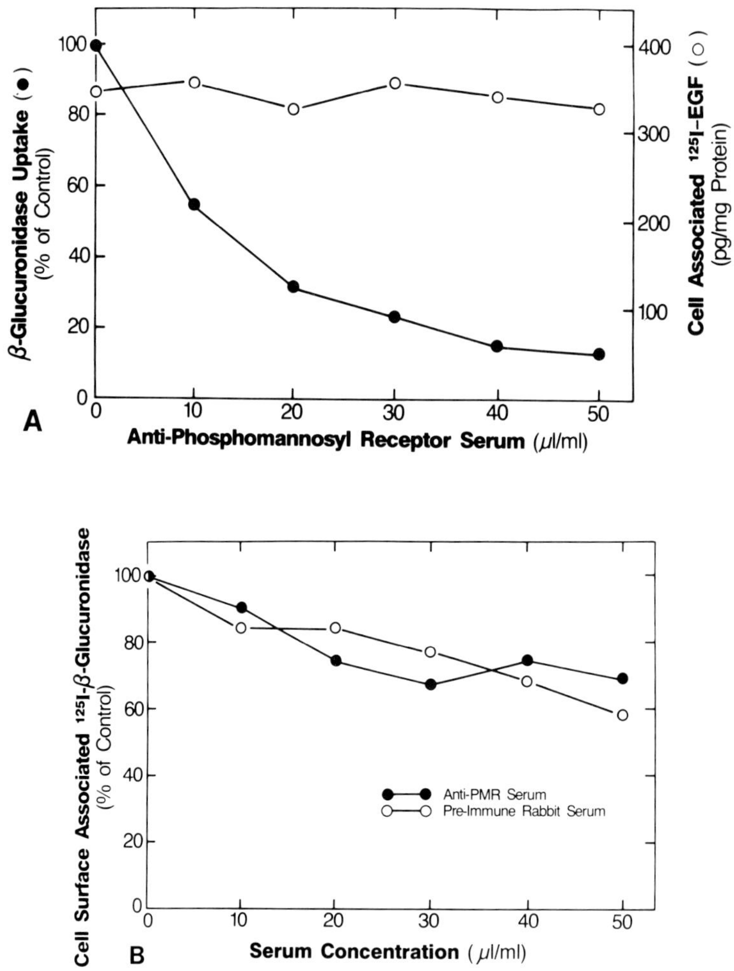

Fig. 1. The effect of antibovine liver-PMR serum on pinocytosis of β-glucuronidase and on cell-surface binding of ^{125}I-β-glucuronidase by human fibroblasts. **A:** Confluent monolayers of β-glucuronidase-deficient fibroblasts, in 35-mm dishes, were incubated for 2 hr, at 37°C, with 1.0 ml of Eagle MEM containing increasing concentrations of anti-PMR serum. The cells were then washed with PBS and pinocytosis of β-glucuronidase (●) and of ^{125}I-EGF (○) were measured. Enzyme pinocytosis refers to M6-P-inhibitible enzyme uptake determined by incubating cells with 4,000 units/ml of human spleen β-glucuronidase for 2 hr at 37°C in the presence or absence of 2 mM M6-P. Enzyme was assayed fluorometrically in cell lysates. Pinocytosis of ^{125}I-EGF was determined following incubation of cells for 30 min at 37°C with 10 μg/ml of ^{125}I-EGF with or without excess unlabeled EGF. Cell lysates were counted for ^{125}I. **B:** Similar monolayers of fibroblasts were incubated for 1 hr at 4°C with 1 ml of Eagle MEM containing the indicated concentrations of normal rabbit serum (○) or anti-PMR serum (●). Following removal of the medium, the cells were washed once with 2 ml of ice-cold medium and M6-P-specific cell-surface binding of ^{125}I-β-glucuronidase was measured. Cells were chilled to 4°C and incubated with ^{125}I-β-glucuronidase (250,000 cpm/4,000 units enzyme activity) for 1 hr at 4°C in the presence or absence of 10 mM M6-P. Cell lysates were counted for ^{125}I. Uptake of β-glucuronidase and ^{125}I-β-glucuronidase binding are expressed as % of control, nonantibody-treated cells, while ^{125}I-EGF uptake is represented as pg/mg protein.

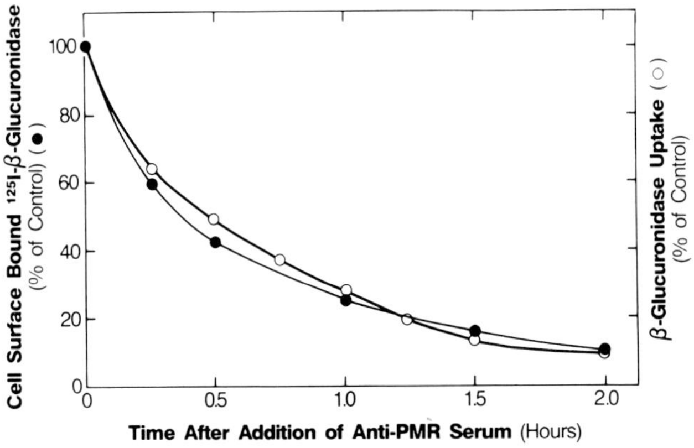

Fig. 2. Kinetics of inactivation of pinocytosis and of loss of cell-surface binding activity. Confluent monolayers of β-glucuronidase-deficient fibroblasts were incubated at 37°C with 1 ml of medium containing 50 μl of anti-PMR serum. At the indicated times, the medium was removed, cells were washed with MEM, and pinocytosis of β-glucuronidase (○) or cell-surface binding of ^{125}I-β-glucuronidase (●) was measured, as described in the legend to Figure 1. Activities are expressed as % of those of control, nonantibody-treated cells.

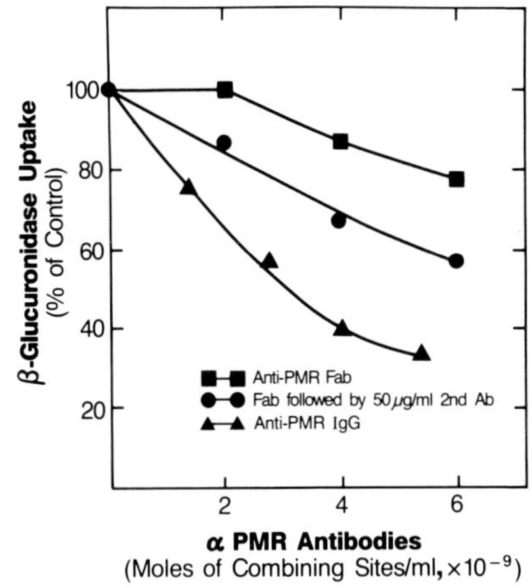

Fig. 3. Cross-linking of phosphomannosyl receptors is of importance in inactivation of β-glucuronidase pinocytosis. Confluent monolayers of β-glucuronidase-deficient fibroblasts, in 35-mm petri dishes, were incubated for 2 hr, at 37°C, with 1 ml of medium containing increasing concentrations of anti-PMR IgG (▲) or of anti-PMR Fab fragments (■). Following exposure to anti-PMR Fab fragments, one set of dishes was incubated for a further 2 hr at 37°C with (50 μg/ml) goat antirabbit IgG (●). The monolayers were then washed with MEM and pinocytosis of β-glucuronidase was measured, as described in the legend to Figure 1. Pinocytotic activity is expressed as % of control nonantibody-treated cells.

Figure 3 shows that anti-PMR Fab fragments inhibited endocytosis by about 25% while intact IgG produced a 65% inhibition. The effectiveness of the Fab fragments could be increased, however, by the addition of a cross-linking second antibody (goat antirabbit IgG, specific for heavy and light chains) to cells previously incubated with Fab fragments (Fig. 3). The second antibody alone had no effect on pinocytosis. These results suggest that cross-linking of receptors by the anti-PMR antibody may be of importance but not necessary in the antibody-induced inactivation of pinocytosis.

We next examined the recovery of cells from antibody treatment. Recovery of pinocytotic activity was slow compared to its inactivation; 48 hr was required for β-glucuronidase uptake of antibody-treated cells to equal 80% that of controls (Fig. 4). This rate of recovery was similar to the turnover rate of PMR in human fibroblasts ($t_{1/2}$ = 24–29 hr) [8] and suggested that synthesis of new receptors is required for recovery of pinocytosis. To test this, antibody-treated cells were allowed to recover in the presence or absence of 100 μM cycloheximide for 24 hr, and cell-surface binding activity was then measured with ^{125}I-β-glucuronidase. In the absence of cycloheximide, antibody-treated cells recovered 72% of the binding of control, nonantibody-treated cells. Cycloheximide-treated cells only reached 25% of normal binding (Fig. 4). (Cells which had not been preincubated with antiserum, but were incubated for 24 hr in the presence of 100 μM cycloheximide, had 63% of control binding activity, consistent with the 24–29-hr half-life of the receptor estimated from metabolic labeling experiments [8].)

The above results suggested that antibody-induced disappearance of cell-surface PMR activity was essentially irreversible in the absence of new protein synthesis. This is in contrast to the chloroquine-induced loss of cell-surface receptors, which is rapidly reversed on removal of the inhibitor [12]. Trypsin treatment of cells also has been reported to cause a loss of cell-surface binding activity which is quite rapidly reversed, at least partially, by externalization of receptors from the intracellular pool [11]. The inability of antibody-treated cells to recover in such a rapid manner suggests that, on prolonged exposure to antibody, the intracellular receptor pool also becomes inactivated by the anti-PMR antibodies. Sahagian has previously shown that intracel-

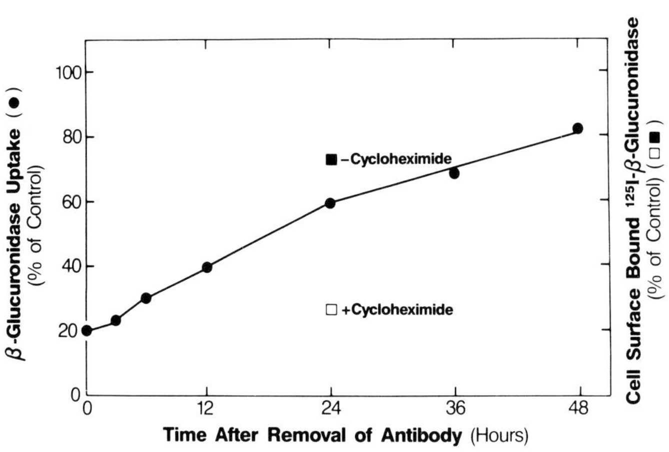

Fig. 4. Recovery of β-glucuronidase pinocytosis and of cell-surface binding activity following removal of anti-PMR serum. Confluent monolayers of β-glucuronidase-deficient fibroblasts, in 35-mm dishes, were incubated with 1 ml of medium containing 50 μl of anti-PMR serum for 2 hr at 37°C. This medium was then removed and cells were incubated at 37°C with nonantibody-containing medium. At the indicated times, β-glucuronidase pinocytotic activity (●) was measured as described in the legend to Figure 1. The effect of cycloheximide on recovery of cell-surface binding activity was examined by incubating cells with medium containing anti-PMR serum for 2 hr, as above. Antibody containing medium was then replaced by fresh medium and cells were allowed to recover in the presence (□) or absence (■) of 100 μM cycloheximide. At the end of the recovery period, cell-surface binding of ^{125}I-β-glucuronidase was measured, as described in the legend to Figure 1. Pinocytotic and binding activities are expressed as % of control plates not exposed to anti-PMR serum.

lular PMR is accessible to exogenously added antibody [31]. It was of interest, therefore, to examine the effect of anti-PMR antibody on the function of the intracellular PMR.

Impairment of intracellular PMR function, by chloroquine or NH_4Cl, results in the enhanced secretion of precursor forms of acid hydrolases [12, 13]. Exposure to anti-PMR antiserum had similar effects. The effects of increasing concentrations of anti-PMR IgG and of normal rabbit IgG on the secretion of β-hexosaminidase by fibroblasts over a 24-hr period are shown in Figure 5. At a concentration of 150 μg/ml of anti-PMR IgG, the level of β-hexosaminidase in the medium was 71% that of cells incubated for 24 hr in the presence of 10 mM NH_4Cl (data not shown). This antibody-induced secretion was not due to nonspecific cell damage or lysis, as indicated by the lack of effects of anti-PMR IgG on extracellular levels of the cytosolic enzyme lactate dehydrogenase (Fig. 5).

The β-hexosaminidase secreted in response to antibody treatment was a "high-uptake" form, as has been previously reported for enzyme precursor secreted in the presence of NH_4Cl [12]. This was indicated by the fact that the rate of endocytosis of this enzyme was equal to that of β-hexosaminidase secreted by NH_4Cl-treated fibroblasts (Table I). Figure 6 shows that β-hexosaminidase B secreted by Tay-Sachs disease fibroblasts during a 21-hr exposure to 150 μg/ml of anti-PMR IgG was the 63-kd precursor form and not the 29-kd processed form which predominates in lysosomes [13]. This indicates that the secreted enzyme was newly synthesized and not enzyme removed from the lysosome. Anti-PMR antibodies, therefore, in addition to inactivating PMR-mediated pinocytosis, disrupt the intracellular sorting of newly synthesized acid hydrolases and enhance their secretion.

It has previously been reported that in cells treated with tunicamycin or chloroquine, the disruption of trafficking of acid hydrolases is correlated with a redistribution of PMR away from its steady-state location [32]. We examined the distribution of PMR in fibroblasts treated with anti-PMR serum, by indirect immunofluorescence.

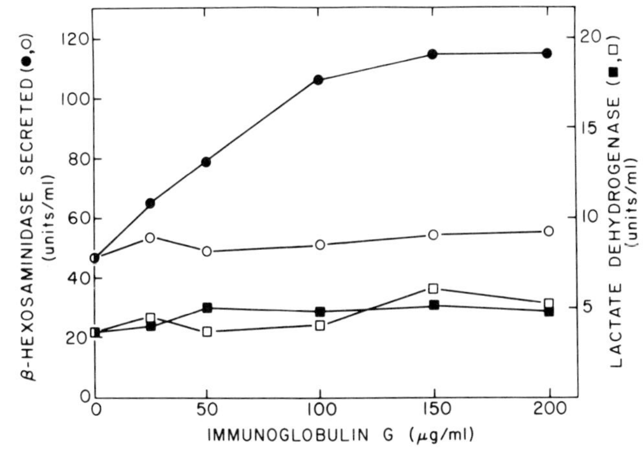

Fig. 5. Effect of anti-PMR IgG on secretion of β-hexosaminidase by human fibroblasts. Confluent 35-mm petri dishes of normal human fibroblasts were incubated for 24 hr in 2.0 ml of Waymouth medium containing 1 mg/ml human serum albumin, 2 mM glutamine, and the indicated concentrations of normal rabbit IgG (○,□), or anti-PMR IgG (●,■). The levels of β-hexosaminidase (○,●) and lactate dehydrogenase (□,■) activities in the media were then determined.

TABLE I. Pinocytosis of β-Hexosaminidase Secreted in Response to Anti-PMR IgG or NH_4C1*

Treatment	Pinocytosis of β-hexosaminidase (units/mg protein/hr)
Anti-PMR IgG (150 μg/ml)	9.5
NH_4C1 (10 mM)	9.1

*Confluent 100-mm petri dishes of normal human fibroblasts were incubated in 5.0 ml of Waymouth medium containing 1 mg/ml human serum albumin, 2 mM glutamine, 2 mM mannose-6-phosphate, and either 150 μg/ml anti-PMR IgG or 10 mM NH_4C1. After 24 hr the medium was removed and the anti-PMR IgG was removed by affinity chromatography on a 0.5 ml column of Protein A-Sepharose (Pharmacia). The medium was dialyzed against PBS to remove mannose-6-phosphate and NH_4C1 and then concentrated tenfold by ultrafiltration using a PM 10 membrane filter (Amicon). Uptake of β-hexosaminidase was measured as described in Materials and Methods.

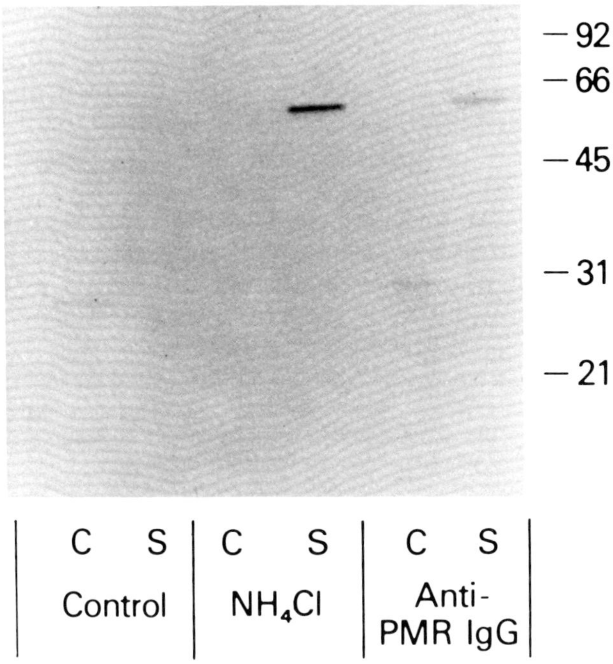

Fig. 6. Anti-PMR IgG causes secretion of precursor β-hexosaminidase B by Tay-Sachs disease fibroblasts. Confluent Tay-Sachs disease fibroblasts were continuously labeled for 21 hr with [2-^{3}H]mannose (60 μCi/ml) in the absence (control) or presence of 10 mM NH_4Cl or 150 μg/ml anti-PMR IgG. Cellular (C) and secreted (S) β-hexosaminidase B was immunoprecipitated and the immunoprecipitates were subjected to SDS-polyacrylamide gel electrophoresis and fluorography, as described in Materials and Methods. The position of the following molecular weight standards are indicated: phosphorylase b, 92,500; bovine serum albumin 66,200; ovalbumin, 45,000; carbonic anhydrase, 31,000; soybean trypsin inhibitor, 21,500. Bio-Rad, Richmond, CA.

Cells which had been preincubated with preimmune rabbit serum for 2 or 8 hr showed a bright, predominantly juxtanuclear fluorescence pattern (Fig. 7A,C), consistent with previously reported localization of the PMR in fibroblasts to the Golgi cisternae, associated coated vesicles, and larger organelles [33]. The cells preincubated with antibovine liver-PMR serum for 2 hr, however, showed a different staining pattern. Staining was much less intense in the juxtanuclear area and was more peripherally located, in a largely punctate pattern (Fig. 7B). Thus, there is an antibody-induced redistribution of receptor which correlates with functional impairment of the PMR's ability to target newly synthesized acid hydrolases to lysosomes. Following pretreatment of cells with anti-PMR serum for 8 hr, the cells did not show any significant staining for PMR (Fig. 7D), suggesting antibody-induced receptor degradation.

The fact that the antibovine liver-PMR antibody inhibited ligand binding by the receptor to only a small extent allowed us to examine the binding activity of the receptor after the antibody-induced redistribution from its normal cell surface and intracellular locations. Cells were incubated with antiserum for various times, following which total cell membranes were prepared and tested for binding of ^{125}I-β-glucuronidase (Fig. 8). Anti-PMR-treated cells showed a gradual decrease in total binding activity (cell surface and intracellular), with a half-life of 7.5 hr. Thus, although loss of enzyme pinocytosis and receptor redistribution occurs rapidly (within 2 hr), the ligand binding property of the internal pool of receptors is much more slowly inactivated. These observations are consistent with the idea that antibody

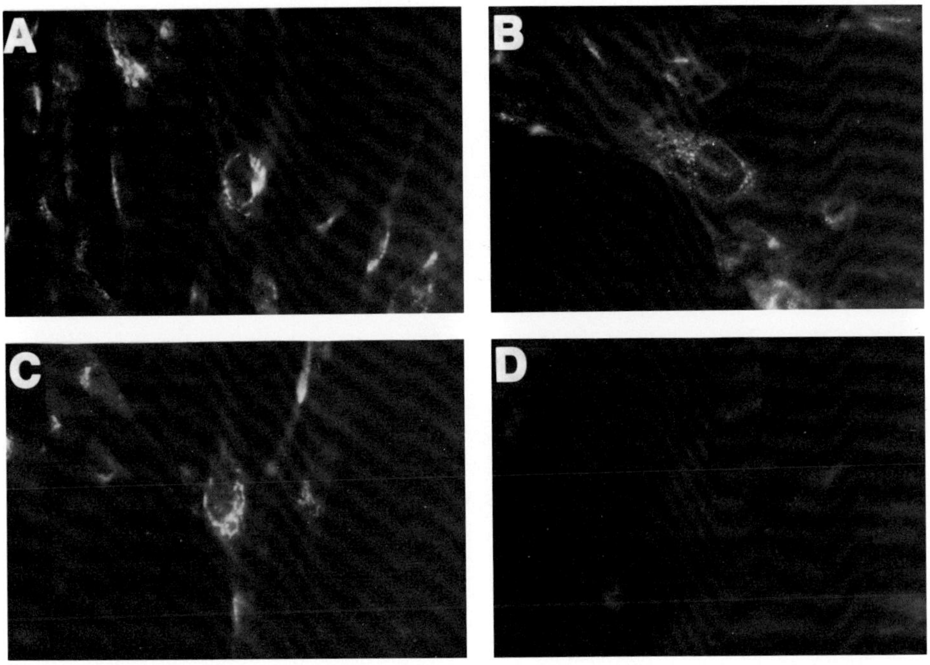

Fig. 7. Anti-PMR serum causes a redistribution of intracellular PMR. Subconfluent monolayers of human fibroblasts were incubated for 2 hr (**A**,**B**) or 8 hr (**C**,**D**) at 37°C, with medium containing a 1:20 dilution of preimmune rabbit serum (A,C) or of antibovine liver PMR serum (B,D). Cells were fixed in methanol and stained for PMR by using goat antihuman liver PMR followed by fluorescein-conjugated $F(ab^1)_2$ rabbit antigoat IgG, as described in Materials and Methods.

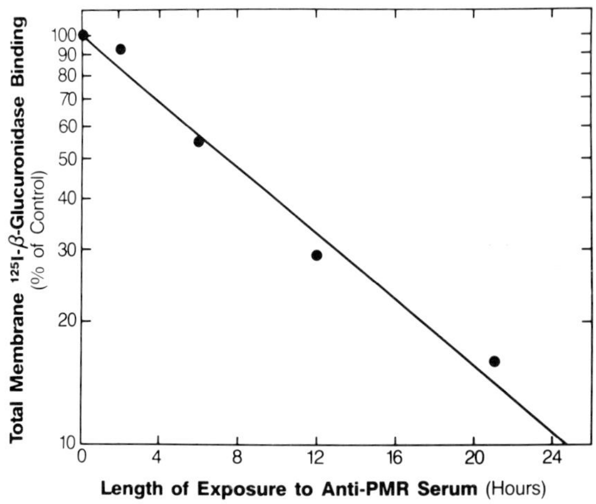

Fig. 8. Kinetics of loss of total membrane binding activity. Confluent monolayers of fibroblasts were incubated with 5 ml of medium containing 250 μl of antibovine liver-PMR serum. At the indicated times, cells were harvested, total membranes were prepared, and binding of ^{125}I-β-glucuronidase was measured by incubating membranes with ^{125}I-β-glucuronidase (250,000 cpm/4,000 units enzyme activity) for 1 hr at 4°C in the presence or absence of 10 mM M6-P. Binding activity is expressed as % of control, nonantibody-treated cells.

treatment inhibits the functioning of the PMR not by blocking binding, but rather by interfering with its recycling to the sites of enzyme sorting (ie, the cell surface for enzyme pinocytosis and the Golgi apparatus for sorting of newly synthesized acid hydrolases).

DISCUSSION

Studies of several cell-surface receptors, including the low-density lipoprotein (LDL) receptor [24], the epidermal growth factor (EGF) receptor [25], transferrin receptor [28], and the insulin receptor [29,30], have shown a reduction in cell-surface receptor number in cells exposed to the respective antibodies. In the case of the LDL [24] and EGF [25] receptors, this reduction has been attributed to the prevention of receptor recycling by antibody-induced cross-linking of receptors. This interpretation was supported by the fact that the effect was produced by polyclonal IgG but not by Fab fragments or by monoclonal antibodies. The loss of cell-surface transferrin-receptor induced by the monoclonal antibody, OKT9, has been suggested to be due to an increased rate of departure of receptors from the recycling pool and not to an inhibition of recycling [28].

Antibody-induced cross-linking of receptors appears to be of importance in the inactivation of endocytosis by anti-PMR antibodies, though this may not be the sole mechanism of inactivation. Monovalent anti-PMR Fab fragments were less inhibitory than intact IgG. The effectiveness of the Fab fragments could be increased by incubating the cells with a second cross-linking antibody which, by itself, had no effect on pinocytosis. It is interesting that anti-PMR IgG was not as effective an inhibitor of endocytosis as the unfractionated antiserum. It may be that the anti-PMR serum contains a population of polyvalent IgM molecules which contribute to the inactivation process and which are lost on preparation of IgG.

While the antibody-induced inhibition of endocytosis and loss of cell-surface receptors are very rapid processes, with half-lives of 25–30 min, full recovery of these activities is much slower and requires synthesis of new receptors. The cell-surface PMR of fibroblasts constitutes only about 10–20% of total cell-associated PMR, the remainder being intracellular [34]. The requirement for synthesis of new receptors for recovery of cell-surface binding activity following antibody treatment suggested that the intracellular PMR pool was also inactivated by the antibody and unable to replace the inactivated cell-surface receptors.

Further evidence that the intracellular pool of enzyme receptors was accessible to and inactivated by anti-PMR antibody was provided by studies of enzyme secretion. Antibody-treated cells showed enhanced secretion of β-hexosaminidase, which was quantitatively similar to the enhanced secretion produced by NH_4Cl, and qualitatively similar in that the secreted form was the precursor enzyme and not the processed mature form.

Whether the intracellular receptors which participate in sorting newly synthesized acid hydrolases are inactivated by antibody at the cell surface, which would imply that they were in equilibrium with the cell-surface receptors which mediate enzyme pinocytosis, or are inactivated by antibody that is internalized by fluid phase endocytosis, is unclear. What is clear is that intracellular PMR is accessible to and functionally inactivated by the anti-PMR antibody. Sahagian has also reported the accessibility of intracellular PMR to exogenously added antibodies and the equilibration of cell-surface and intracellular pools of receptor [31].

We have previously proposed that PMR function, both in enzyme pinocytosis and in sorting newly synthesized acid hydrolases, is dependent on receptor reutilization or recycling [12]. We have suggested that the impairment of both processes by amines is explainable by the ability of these agents to inhibit receptor recycling. Similarly, we propose that the inhibition of both of these processes by antibody which does not block ligand binding is due to impairment of receptor recycling. However, unlike impairment by amines, which is secondary to inhibition of pH-dependent dissociation of ligand from receptor, and which is freely reversible on removal of amines, inactivation of receptors by antibodies is essentially irreversible. There appear to be two components to the inactivation—a removal of functional receptor from the recycling pool, which is rapid and which we attribute to receptor cross-linking, and a slower process of somewhat accelerated degradation of antibody-inactivated receptor, the mechanism of which is not yet clear. The morphological equivalent of the early component of the inactivation is redistribution of most of the receptors from perinuclear (presumably Golgi) distribution to more peripheral vesicles. The morphological equivalent of the second component of the inactivation is disappearance of the immunofluorescent PMR signal.

Prior studies have suggested that most of the normal turnover of the PMR in fibroblasts does not involve degradation in lysosomes, since turnover was nearly normal in I-cell fibroblasts and was not markedly inhibited by NH_4Cl or leupeptin [8]. We are also unable to demonstrate any effect of leupeptin or NH_4Cl on the rate of loss of binding activity in antibody-treated cells (data not shown). Thus, the site of enhanced turnover of the PMR is not clear.

Antibody-induced PMR deficiency in human fibroblasts has previously been reported by von Figura and co-workers [7,35]. This deficiency was also characterized by impairment of endocytosis of exogenous ligands and of targeting of newly synthesized acid hydrolases. However, the mechanism of PMR inactivation appears to be different. This may be due to differences in the antibodies used. The antihuman liver-PMR antibodies used by von Figura's group blocked the PMR binding site, and inhibition of pinocytosis and targeting was attributed to this property. Furthermore, these authors concluded that neither polyvalent nor monovalent antibodies prevented receptor cycling between the cell surface and the intracellular compartments. This interpretation contrasts with the conclusions we draw from the results reported here. With our antibody, inactivation cannot be readily explained by interference with ligand binding. We concluded that cross-linking of receptors by antibody is of importance in the inactivation of receptor by this antibody, and that the inactivation results from the inability of cross-linked receptor which is still able to bind ligand, to recycle to the site of sorting.

Several conditions have now been established in which function of phosphomannosyl receptors is impaired. These include conditions where the ligand is defective, as in I-cell fibroblasts [36] or in normal cells treated with tunicamycin [37], and conditions where ligand-receptor dissociation is prevented, as in NH_4Cl- or chloroquine-treated cells [12,13], or in mutant cells which are defective in the ability to acidify their endosomes [38]. The antibody-induced impairment of PMR function reported in this manuscript and that observed by von Figura and co-workers [7,35] represent systems in which the disruption of receptor function is due to direct effects on the receptor itself, in one case blocking the ligand binding to receptor, and in the other, blocking receptor recycling. The ability to perturb the function of the PMR with these reagents may be of use in studying the pathways by which acid hydrolases are sorted.

ACKNOWLEDGMENTS

This work was supported by grants from the USPHS, NIH GM34182 and AI22359, and from the National Foundation, March of Dimes.

We would like to thank Dr. C. Tsai, Pathology Department, St. Louis University, for his help with the immunofluorescence studies.

REFERENCES

1. Neufeld EF, Ashwell G: In Lennarz WJ (ed): "The Biochemistry of Glycoproteins and Proteoglycans." New York: Plenum Press, 1980, pp 241–266.
2. Sly WS, Fischer HD: J Cell Biochem 18:67–85, 1982.
3. Fischer HD, Creek KE, Sly WS: J Biol Chem 257:9938–9943, 1982.
4. Sahagian GG, Distler J, Jourdian GW: Proc Natl Acad Sci USA 78:4289–4293, 1981.
5. Steiner AW, Rome LH: Arch Biochem Biophys 214:681–687, 1982.
6. Geuze HJ, Slot WJ, Strous GJAM, Hasilik A, von Figura K: J Cell Biol 98:2047–2054, 1984.
7. von Figura K, Gieselmann V, Hasilik A: EMBO J 3:1281–1286, 1984.
8. Creek KE, Sly WS: Biochem J 214:353–360, 1983.
9. Sahagian GG, Distler J, Jourdian GW: Methods Enzymol 83:392–396, 1982.
10. Sahagian GG, Neufeld EF: J Biol Chem 257:7121–7128, 1983.
11. Rome LH, Weissman B, Neufeld EF: Proc Natl Acad Sci USA 76:2331–2334, 1979.
12. Gonzalez-Noriega A, Grubb JH, Talkad V, Sly WS: J Cell Biol 85:839–852, 1980.
13. Hasilik A, Neufeld EF: J Biol Chem 255:4937–4945, 1980.
14. Natowicz MR, Chi MM-Y, Lowry OH, Sly WS: Proc Natl Acad Sci USA 76:4322–4326, 1979.
15. Talkad V, Sly WS: J Biol Chem 258:7345–7351, 1983.

16. Good AH, Wofsy L, Kimura J, Henry C: In Mishell B, Shiigi S (eds): "Selected Methods in Cellular Immunology." San Francisco: W.H. Freeman & Co., 1980, pp 278–286.
17. Mage MG: Methods Enzymol 70:142–150, 1980.
18. Laemmli UK: Nature 227:680–685, 1970.
19. Fischer HD, Gonzalez-Noriega A, Sly WS: J Biol Chem 255:5069–5074, 1980a.
20. Fischer HD, Gonzalez-Noriega A, Sly WS, Morre DJ: J Biol Chem 255:9608–9615, 1980b.
21. Reeves WJ, Fimognari GM: Methods Enzymol 9:288–294, 1986.
22. Lowry OH, Rosebrough NJ, Farr AL, Randall RJ: J Biol Chem 193:265–275, 1951.
23. Kaplan A, Achord DT, Sly WS: Proc Natl Acad Sci USA 79:2026–2030, 1977.
24. Anderson RGW, Brown MS, Beisiegal V, Goldstein JL: J Cell Biol 93:523–531, 1982.
25. Schreiber AB, Lieberman TA, Lax I, Yarden Y, Schlessinger J: J Biol Chem 258:846–854, 1983.
26. Mellman I, Plutner H, Ukkonen P: J Cell Biol 98:1163–1169, 1984.
27. Mellman I, Plutner H: J Cell Biol 98:1170–1177, 1984.
28. Weissman AM, Klausner RD, Rao K, Harford JB: J Cell Biol 102:951–958, 1986.
29. Roth RA, Maddux BA, Cassell DJ, Goldfine ID: J Biol Chem 258:12094–12097, 1983.
30. Grunfeld C: Proc Natl Acad Sci USA 81:2508–2511, 1984.
31. Sahagian GG: Biol Cell 51:207–214, 1984.
32. Brown WJ, Constantinescu E, Farquhar MG: J Cell Biol 99:320–326, 1984.
33. Brown WJ, Farquhar MG: Proc Natl Acad Sci USA 81:5135–5139, 1984.
34. Fischer HD, Gonzalez-Noriega A, Sly WS: J Biol Chem 255:5069–5074, 1980.
35. Gartung C, Braulke T, Hasilik A, von Figura K: EMBO J 4:1725–1730, 1985.
36. Hickman S, Neufeld EF: Biochem Biophys Res Commun 49:992–999, 1972.
37. von Figura K, Rey M, Prinz R, Voss B, Ullrich K: Eur J Biochem 101:103–109, 1979.
38. Robbins AR, Peng SS, Marshall JL: J Cell Biol 96:1064–1071, 1983.

Journal of Cellular Biochemistry 36:51–58 (1988)
Molecular Biology of Intracellular Protein Sorting and Organelle Assembly 25–32

Lipid Polarity and Sorting in Epithelial Cells

Gerrit van Meer and Kai Simons

European Molecular Biology Laboratory, D-6900 Heidelberg, Federal Republic of Germany

Apical and basolateral membrane domains of epithelial cell plasma membranes possess unique lipid compositions. The tight junction, the structure separating the two domains, forms a diffusion barrier for membrane components and thereby prevents intermixing of the two sets of lipids. The barrier apparently resides in the outer, exoplasmic leaflet of the plasma membrane bilayer. First data are now available on the generation of these differences in Madin-Darby canine kidney (MDCK) cells, grown on filter supports. Experiments in which fluorescent precursors of apical lipids were introduced into the cell have demonstrated that upon biosynthesis apical lipids are sorted from basolateral lipids in an intracellular compartment. In this paper we present a model for the sorting process, the central point of which is that the two sets of lipids laterally segregate into microdomains that bud to form vesicles delivering the lipids to the apical and the basolateral plasma membrane domains, respectively.

Key words: **lipid phases, sphingolipids, Golgi complex, plasma membrane polarity, membrane domain formation**

The plasma membrane of epithelial cells is differentiated into two domains, the apical membrane lining the epithelial lumen and the basolateral domain which faces the blood supply of the tissue [1]. The two plasma membrane domains are responsible for the vectorial functions which characterize transporting, absorptive, and secretory epithelia. The epithelial cells accomplish their polarized functions by localizing a distinct set of cell surface components to either of the two plasma domains. Not only do the apical and the basolateral cell surfaces have distinct protein compositions (enzymes, transport proteins, channels, receptors, etc), but they also contain different lipids. The most complete analysis of the lipid compositions has been reported for intestinal cells [2–4]. The major differences in lipid composition are a 2–4-fold enrichment of glycosphingolipids and a 2–4-fold depletion of phosphatidylcholine (PC) in the apical membrane. The amount of the second major phospholipid, phosphatidylethanolamine (PE), as a percentage of total lipid is about equal in the two domains. In other cell types apical enrichment of the sphingophospholipid, sphingo-

G. van Meer's present address is Department of Cell Biology, Medical Faculty, State University of Utrecht, Nicolaas Beetsstraat 22, 3511 HG Utrecht, The Netherlands.

Received May 19, 1987; accepted June 16, 1987.

myelin, is also observed [eg, 5,6]. In Madin-Darby canine kidney (MDCK) cells the phospholipid differences between the domains are very similar to those observed in intestinal cells [7,8]. MDCK cells also have a very high content of glycosphingolipids, which, as in intestinal cells, are enriched to the apical domain [9–12]. How the polarized distribution of cell surface lipids is generated and maintained in MDCK cells is the subject of this short review.

TIGHT JUNCTIONS MAINTAIN LIPID POLARITY

The boundary between the two cell surface domains is formed by a specific zone of cell-cell contact, which encircles the top of each cell [13]. This structure connects neighbouring cells and seals the spaces between the cells. The sealing function has been well characterized [14]. In addition, the tight junction has been suggested to have a role in maintaining the compositional differences between the two domains by forming a diffusion barrier in the plane of the membrane [14]. Very little is known about the molecular arrangement at the membrane contact sites of the tight junction. Models for the structure of the tight junction have been devised on the basis of its ultrastructural features [15–17]. At the site of the tight junction the intercellular space is occluded, and the two apposed plasma membranes seem to have fused or merged. In freeze-fracture replicas, a complementary pattern of anastomosing strands and furrows is observed, which reflects the molecular organization of the junctional sites. The peculiarities of the strand structures and the membrane fusion images led Kachar, Pinto da Silva, and Reese to postulate that the strands of the tight junctions represent cylindrical, inverted lipid micelles [16,17]. In this lipid model of the zonula occludens, the cytoplasmic leaflets of the apical and the basolateral plasma membrane domains are continuous, while the exoplasmic leaflets are interrupted by the hexagonal lipid cylinder. The model also postulates continuity between the exoplasmic leaflets of the plasma membrane domains of adjacent cells.

Based on the lipid model of the tight junction, interesting predictions concerning the diffusion of membrane lipids can be made. Passage of lipids between the apical and the basolateral domains would depend on their orientation in the bilayer. They would diffuse within the continuous cytoplasmic leaflet from one domain of the cell surface to the other, whereas in the exoplasmic bilayer leaflet the tight junction would act as a barrier between the apical and the basolateral domains. The model further predicts that lipid molecules in the exoplasmic leaflet would be able to diffuse through a continuous lipid monolayer from one cell to another. A study using fluorescent lipid probes indeed suggested that movement from the apical to the basolateral side took place [18]. However, the use of water-soluble lipids in this study left open the possibility that movement occurred via the cytoplasmic aqueous phase.

We have tried to overcome this problem by a more direct experimental approach to the question of tight junction permeability [8]. First of all, we chose an experimental system that has been carefully studied for basolateral and apical protein polarity: two strains of MDCK cells grown on permeable supports [1]. By growing MDCK cells on filters, conditions are achieved that closely mimic those prevailing in vivo. On filters, the cells feed from the basal side and form a monolayer of cuboidal cells, which is more stable to experimental manipulations than cells grown on solid supports. Second, we developed methods to fuse liposomes with the apical plasma membrane domain [19,20]. Because of this we were able to use a fluorescent phospholipid, N-rhodamine-dioleoyl PE (N-Rh-PE), that in its water-insolubility is more natural than the analogs previously used. The fusion reaction also made it possible to insert the fluorescent lipid into both leaflets of the plasma membrane bilayer by the use of symmetric liposomes, or primarily into the exoplasmic leaflet by the use of liposomes in which the fluorescent lipid was enriched in the external leaflet. Third, the behavior of the fluorescent lipid was studied at 0°C, at which temperature diffusion should occur but vesicular transport is completely inhibited. The fusion reaction took place within 1 min at 37°C, a time which is too short for transcytosis to play a role in the transport to the basolateral surface [21,22]. Intactness of the tight junction was assessed by measuring the trans-epithelial electrical resistance [23]. Using these methods, we were able to observe clear differences in the behavior of implanted N-Rh-PE in MDCK cells [8]. When fused into the exoplasmic leaflet of the apical plasma membrane, N-Rh-PE did not distribute to the lateral plasma membrane unless the junctions were opened by calcium removal. However, when N-Rh-PE was fused into both the exoplasmic and cytoplasmic leaflets of the bilayer, a fraction of the fluorescent phospholipid immediately passed to the basolateral plasma membrane domain. Also glycolipids have been shown to be unable to pass the tight junction in the exoplasmic bilayer leaflet [24]. These results agree with the predictions of the hexagonal lipid model for tight junction structure.

We also tested whether lipids can diffuse from one epithelial cell to another through continuous exoplasmic leaflets of their apical plasma membranes by studying the behavior of endogenous glycolipids [25]. Strain II MDCK cells, a subline of the MDCK cell line, possess a series of glycolipids, the globo series, which are not found in MDCK strain I cells, a different MDCK subline [10]. Forssman antigen, one glycolipid of this series [GalNAc(α1-3)GalNAc(β1-3)Gal(α1-4)Gal(β1-4)Glc(β1-1)Cer], constitutes 21% of the total neutral glycosphingolipids of MDCK strain II cells. When a monolayer of these cells was labeled from the apical side with a monoclonal antibody against Forssman antigen, apical staining was observed [25]. This glycolipid was therefore present in the exoplasmic leaflet of the apical plasma membrane. We co-cultured MDCK strain II cells with MDCK strain I cells which do not express Forssman antigen. The presence of intact tight junctions between the two cell types was demonstrated by adding a monoclonal antibody which recognizes uvomorulin, a protein present on the lateral membrane of MDCK cells [26], and showing that it did not label the lateral membrane unless the tight junctions had been opened by EGTA treatment before fixation. Therefore, tight junctions were present between co-cultured strain I and II cells. Staining with a monoclonal antibody against Forssman antigen demonstrated the presence of antigen on the apical surface of about 50% of the cells in the mixed monolayer. The boundary between stained and unstained cells was sharp. Thus, the endogenous glycolipid could not pass from MDCK strain II cells to the apical surface of neighboring MDCK strain I cells over a period of 96 hr of coculture at 37°C.

These observations argue against a continuity between the external leaflets of the apical plasma membrane in adjacent epithelial cells. The results are thus difficult to reconcile with the hexagonal lipid model of tight junction structure where the outer leaflets of the plasma membranes of neighboring cells are fused. It seems more likely that proteins which are part of the tight junction structure [27] bring the plasma membranes of adjacent cells very close together but do not induce a partial fusion. This would also be more consistent with the fact that the tight junction, as a barrier to ion diffusion between cells, displays ion selectivity [14].

If one assumes that the free diffusion of lipid molecules in the cytoplasmic leaflet of the plasma membrane leads to an identical lipid composition of the cytoplasmic leaflets of the apical and basolateral plasma membrane domains, interesting consequences arise for the transbilayer distribution of the individual lipid classes in the apical and basolateral membranes. If the exoplasmic leaflet of the apical domain were predominantly occupied by glycosphingolipids, as is probably the case in intestinal cells [see 2–4,8], the phospholipids of the apical domain would be mainly situated in its cytoplasmic leaflet. The phospholipid composition of the cytoplasmic leaflet of the basolateral membrane would be identical with that of the apical domain, and the distribution of the individual phospholipid classes across the basolateral membrane bilayer could be predicted from the total phospholipid composition of the basolateral membrane. For the two major phospholipid classes this leads to the following predictions: 65–90% of the PE and only 10–25% of the PC would be localized in the cytoplasmic leaflet [3,7,8]. This agrees with the distribution reported for the plasma membrane of erythrocytes and blood platelets [28,29], and may reflect general principles underlying phospholipid organization in mammalian plasma membranes.

SORTING OF PLASMA MEMBRANE LIPIDS IN THE GOLGI COMPLEX

These observations suggest that the lipid differences between the apical and the basolateral membranes reside in their exoplasmic leaflets. The problem of generating the different lipid compositions of these two membrane domains is reduced to enriching sphingolipids, predominantly glycosphingolipids, and PC in the exoplasmic leaflets of the apical and the basolateral membranes respectively. How are these differences generated? Several possibilities exist. One is local metabolism. This seems very unlikely since PC is synthesized mainly in the endoplasmic reticulum, and glycosphingolipids in the Golgi complex [30,31]. These lipids would therefore have to be transported to the cell surface from their sites of synthesis. In principle, this could occur by two routes, either through the cytosol—eg, by the help of exchange proteins—or by vesicular transport. If vesicular transport were the mechanism for distributing the correct lipids apically and basolaterally, one would expect that sorting would occur in the membrane compartment from which vesicles exit for the cell surface. This is the trans-most compartment of the Golgi complex, the trans-Golgi network [32]. This is also the site where newly synthesized apical and basolateral proteins are thought to be sorted from each other before their transport to the cell surface [33].

In a recent study [34] we addressed the question of whether lipid sorting occurs in the Golgi complex by incorporating into MDCK fluorescent precursor of glycosphingolipids and sphingomyelin, allowing their fluorescent products to accumulate in the membrane of the Golgi complex and monitoring their subsequent delivery to the apical and basolateral cell surface. We used the fluorescent ceramide analog N-6[7-nitro-2,1,3,-benzoxadiazol-4-yl]amino-caproyl sphingosine (C6-NBD-ceramide) introduced by Lipsky and Pagano [35]. This probe partitions into the membranes of tissue culture cells at 2°C and becomes trapped in the Golgi complex upon conversion to C6-NBD-sphingomyelin and C6-NBD-glucosylceramide at 20°C (Fig. 1). These products then move to the plasma membrane at 37°C from where they can be depleted by means of a "back-exchange" reaction with unlabeled liposomes. The first conclu-

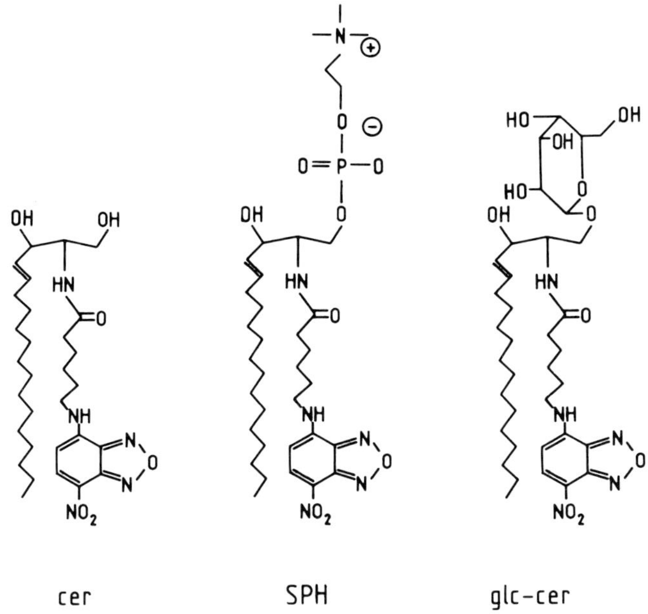

Fig. 1. The C6-NBD ceramide is metabolized to C6-NBD-sphingomyelin and C6-NBD-glucosylceramide in the Golgi complex [35].

sion of our study was that the fluorescent sphingolipids, C6-NBD-sphingomyelin and C6-NBD-glucosylceramide, were selectively incorporated into the luminal leaflet of the membrane of the Golgi complex. Secondly, both fluorescent lipids were delivered to the cell surface by vesicular transport and, most importantly, the cell surface delivery was polar: per unit of surface area, roughly ten times more C6-NBD-glucosylceramide and three times more C6-NBD-sphingomyelin were delivered to the apical plasma membrane domain. This enrichment was due to direct delivery to the two surface domains and not a consequence of plasma membrane recycling, since the experiments were performed under conditions in which both domains were continually depleted of newly delivered fluorescent lipid. Sorting must therefore have occurred before the fluorescent lipids reached the plasma membrane.

These results suggest that the cell surface polarity of lipids may be generated by lipid sorting in the trans-Golgi network. The topology of the bilayer differences is compatible with the physical separation of sphingolipids and PC in the luminal leaflet of the Golgi membrane, subsequent budding of the sphingolipid domain into a membrane vesicle, and fusion of the vesicle with the apical membrane introducing the sphingolipids into the exoplasmic leaflet of the apical lipid bilayer (Fig. 2). According to this working hypothesis, the specific event in the lipid sorting process is the generation of the sphingo(glyco)lipid micro-domain in the luminal leaflet of the trans-Golgi network. The luminal leaflet of the membrane which buds into basolateral transport vesicles would, as a consequence, be depleted of the sphingo(glyco)lipids and enriched in PC. This view assumes that the phospholipid asymmetry across the bilayer has been generated earlier in the pathway, possibly already in the membrane of the endoplasmic reticulum [36], and that the asymmetric distribution of choline and amino-containing phospholipids is maintained during the membrane vesiculations

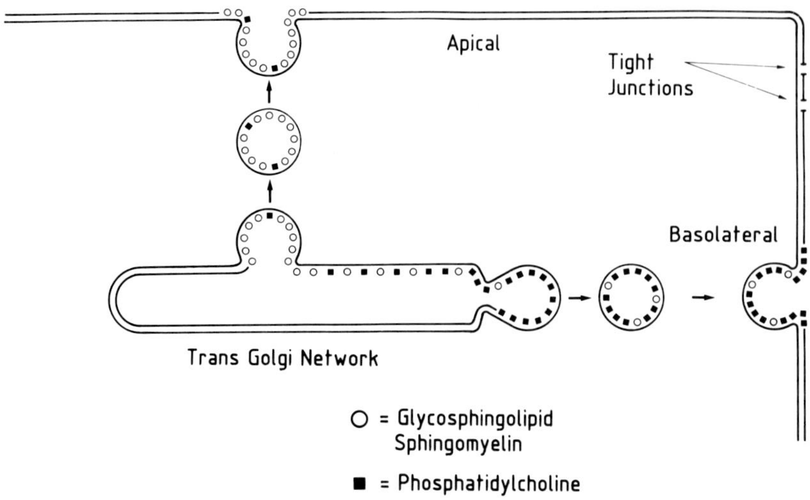

Fig. 2. Sorting of apical and basolateral lipids by lateral segregation in the luminal bilayer leaflet of the trans-Golgi network.

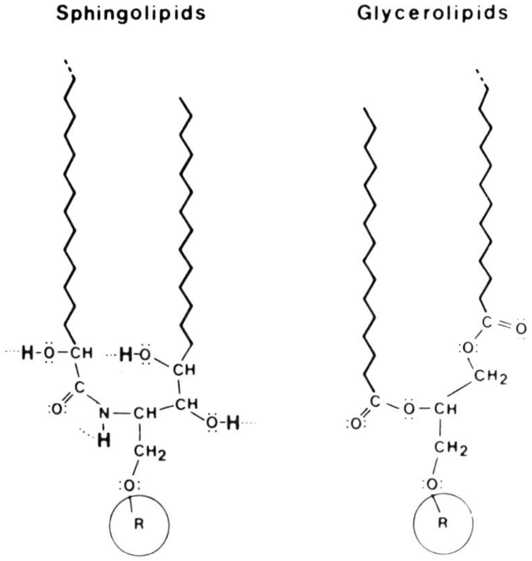

Fig. 3. Differences in hydrogen bond donors and acceptors in the ceramide part of sphingolipids and the diglyceride part of glycerolipids. (Reproduced from Pascher [37] with permission.)

that occur in the biosynthetic transport pathway starting in the endoplasmic reticulum and ending at the cell surface.

Our working hypothesis for lipid sorting in the Golgi complex assumes the formation of a microdomain of sphingo(glyco)lipids in the luminal leaflet of the bilayer. If this were true, this would be the first clear example where the formation of microdomains of lipids is of functional importance for the cell. One interesting feature of the process is that the sphingolipids have a common structural feature, their sphingosine backbone which is not present in glycerolipids. These sphingolipids, especially the glycosphingolipids, have a tendency to associate by hydrogen bond formation to form separate domains in model lipid systems [37,38]. This domain formation could be facilitated by the conditions prevailing in the lumen of the intracellular sorting compartment (low pH, ionic conditions, etc.). Glycerolipids cannot form intermolecular hydrogen bonds in the same regions of the lipid bilayer since they only have groups which can act as hydrogen bond acceptors (Fig. 3).

Until the putative transport vesicles carrying sphingolipids to the apical membrane have been isolated, we will not know whether these also are responsible for transporting apical proteins to their destination. The delivery of apical lipids and proteins show the same kinetics, and are both slowed down by the 20°C temperature block. It is thus hard to imagine that the two sorting processes would not be related. The enticing possibility exists that specific sorting in membrane traffic involves both proteins and lipids interacting with each other in a cooperative fashion to form a lattice which buds into a vesicle. The sphingolipids could bind to sorting proteins and/or apical proteins during microdomain formation in the trans-Golgi network. This interaction would have to be reversed upon arrival at the apical surface domain (change of pH, ionic conditions, etc) to allow the putative sorting proteins to return to the trans-Golgi network to repeat their sorting functions.

ACKNOWLEDGMENTS

We are grateful to Thomas Gabran and Hilkka Virta for the excellent technical assistance in the experimental work reviewed here, and to Mark Bennett for a critical reading of the manuscript.

REFERENCES

1. Simons K, Fuller SD: Annu Rev Cell Biol 1:243–288, 1985.
2. Forstner GG, Wherrett JR: Biochim Biophys Acta 306:446–459, 1973.
3. Kawai K, Fujita M, Nakao M: Biochim Biophys Acta 369:222–233, 1974.
4. Brasitus TA, Schachter D: Biochemistry 19:2763–2769, 1980.
5. Meier PJ, Sztul ES, Reuben A, Boyer JL: J Cell Biol 98:991–1000, 1984.
6. Molitoris BA, Simon FR: J Membr Biol 83:207–215, 1985.
7. van Meer G, Simons K: EMBO J 1:847–852, 1982.
8. van Meer G, Simons K: EMBO J 5:1455–1464, 1986.
9. Hougland AE, Gaush C, Mavberry WR: Biochem J 177:311–317, 1979.
10. Hansson GC, Simons K, van Meer G: EMBO J 5:483–489, 1986.
11. Nichols GE, Lovejoy JC, Borgman CA, Sanders JM, Young Jr, WW: Biochim Biophys Acta 887:1–12, 1986.
12. Nimura Y, Ishizuka I: J Biochem (Tokyo) 100:825–835, 1986.
13. Farquhar MG, Palade GE: J Cell Biol 17:375–412, 1963.
14. Diamond JM: Physiologist 20:10–18, 1977.
15. Hirokawa N: J Ultrastruct Res 80:288–301, 1982.
16. Pinto da Silva P, Kachar B: Cell 28:441–450, 1982.
17. Kachar B, Reese TS: Nature 296:464–466, 1982.
18. Dragsten PR, Blumenthal R, Handler JS: Nature 294:718–722, 1981.
19. van Meer G, Simons K: J Cell Biol 97:1365–1374, 1983.
20. van Meer G, Davoust J, Simons K: Biochemistry 24:3593–3602, 1985.
21. Pesonen M, Ansorge W, Simons K: J Cell Biol 99:796–802, 1984.
22. von Bonsdorff C-H, Fuller SD, Simons K: EMBO J 4:2781–2792, 1985.
23. Cereijido M, Robbins ES, Dolan WJ, Rotunno CA, Sabatini DD: J Cell Biol 77:853–880, 1978.
24. Spiegel S, Blumenthal R, Fishman PH, Handler JS: Biochim Biophys Acta 821:310–318, 1985.
25. van Meer G, Gumbiner B, Simons K: Nature 322:639–641, 1986.
26. Gumbiner B, Simons K: J Cell Biol 102:457–468, 1986.
27. Stevenson BR, Siliciano JD, Mooseker MS, Goodenough DA: J Cell Biol 103:755–766, 1986.
28. Bretscher MS: J Mol Biol 71:523–528, 1972.
29. Op den Kamp JAF: Annu Rev Biochem 48:47–71, 1979.
30. Bell RM, Ballas LM, Coleman RA: J Lipid Res 22:391–403, 1981.

31. Pagano RE, Sleight RG: Trends Biochem Sci 10:421–425, 1985.
32. Griffiths G, Simons K: Science 234:438–443, 1986.
33. Hughson E, Wandinger-Ness A, Gausepohl H, Griffiths G, Simons K: In Schwartz M (ed): "The Centenary Symposium of the Pasteur Institute", (in press).
34. van Meer G, Stelzer EHK, Wijnaendts van Resandt RW, Simons K: J Cell Biol 105:1623–1635, 1987.
35. Lipsky NG, Pagano RE: Science 228:745–747, 1985.
36. Bishop WR, Bell RM: Cell 42:51–60, 1985.
37. Pascher I: Biochim Biophys Acta 455:433–451, 1976.
38. Thompson TE, Barenholz Y, Brown RE, Correa-Freire M, Young WW, Tillack TW: In Freysz L (ed): "Enzymes of Lipid Metabolism." New York: Plenum Press, 1986, pp 387–396.

Journal of Cellular Biochemistry 36:59–71 (1988)
Molecular Biology of Intracellular Protein Sorting and Organelle Assembly 33–45

Transport of Proteins Into Yeast Mitochondria

Adolphus P.G.M. van Loon, Martin Eilers, Alison Baker, and Keith Verner

Department of Biochemistry, Biocenter, University of Basel, CH-4056 Basel, Switzerland

The amino-terminal sequences of several imported mitochondrial precursor proteins have been shown to contain all the information required for transport to and sorting within mitochondria. Proteins transported into the matrix contain a matrix-targeting sequence. Proteins destined for other submitochondrial compartments contain, in addition, an intramitochondrial sorting sequence. The sorting sequence in the cytochrome c_1 presequence is a stop-transport sequence for the inner mitochondrial membrane. Proteins containing cleavable presequences can reach the intermembrane space by either of two pathways: (1) Part of the presequence is transported into the matrix; the attached protein, however, is transported across the outer but not the inner membrane (eg, the cytochrome c_1 presequence). (2) The precursor is first transported into the matrix; part of the presequence is then removed, and the protein is reexported across the inner membrane (eg, the precursor of the iron-sulphur protein of the cytochrome bc_1 complex).

Matrix-targeting sequences lack primary amino acid sequence homology, but they share structural characteristics. Many DNA sequences in a genome can potentially encode a matrix-targeting sequence. These sequences become active if positioned upstream of a protein coding sequence. Artificial matrix-targeting sequences include synthetic presequences consisting of only a few different amino acids, a known amphiphilic helix found inside a cytosolic protein, and the presequence of an imported chloroplast protein.

Transport of proteins across mitochrondrial membranes requires a membrane potential, ATP, and a 45-kd protein of the mitochondrial outer membrane. The ATP requirement for import is correlated with a stable structure in the imported precursor molecule. We suggest that transmembrane transport of a stably folded precursor requires an ATP-dependent unfolding of the precursor protein.

Key words: origin of presequences, intracellular and intramitochondrial protein sorting, stop-transport sequence, ATP requirement, protein unfolding, translocation machinery

How are proteins sorted between the various intracellular and intramitochondrial compartments? How are proteins transported across membranes? Is there a translo-

Adolphus P.G.M. van Loon's present address is Dept of Biochemistry, Central Research Unit (ZFE), Hoffmann-La-Roche & Co. AG, Granzacherstrasse 124, CH-4002 Basel, Switzerland.

Received April 6, 1987; revised and accepted July 7, 1987.

cation machinery and what are its components? What are the characteristics of presequences and what is their evolutionary origin? Much progress has been made in the last few years in answering these questions. Here we review some recent results concerning transport of proteins into mitochondria.

PRESEQUENCES CAN TRANSPORT PROTEINS TO AND INTO MITOCHONDRIA

Most mitochondrial proteins are encoded in the nucleus. They are synthesized in the cytoplasm and can be imported into mitochondria post-translationally. Most imported proteins are initially made as precursor molecules with amino terminal extensions. These presequences are typically about 20–30 amino acids long but may be as long as 80 amino acids. Proteins of the mitochondrial outer membrane are synthesized without cleavable presequences [reviewed in 1–3].

Gene-fusion experiments have shown that several cleavable presequences, and also the noncleavable amino terminus of an outer membrane protein, contain sufficient information for transporting attached "passenger" proteins to mitochondria and into the proper intramitochondrial space. Conversely, mitochondrial proteins without their presequences are not transported to mitochondria [reviewed in 4]. Presequences thus can contain information for intracellular targeting to mitochondria and for intramitochondrial sorting. Presequences can also determine the intracellular sorting of naturally occurring isoenzymes: a cytosolic isoenzyme of alcohol dehydrogenase (ADH II) was imported into the mitochondrial matrix if it was fused to the cleavable presequence of the mitochondrial isoenzyme ADHIII, while ADHIII without its presequence remained in the cytosol [5]. Thus, presequences are essential for transporting authentic mitochondrial proteins to mitochondria; they can transport foreign proteins specifically to mitochondria, and they can determine the intracellular sorting of naturally occurring isoenzymes.

MATRIX-TARGETING SEQUENCES CAN BE CONSTRUCTED BY USING ONLY THREE DIFFERENT AMINO ACIDS, PROVIDED THAT THE RESULTING PRESEQUENCE IS AMPHIPHILIC

Presequences of mitochondrial matrix proteins (matrix-targeting sequences) are generally rich in basic and hydroxylated amino acids and lack acidic residues. They have no apparent homology in their primary amino acid sequences but share several structural characteristics [3,4] (Table I). The entire presequence is not required for intracellular targeting to the mitochondrial matrix. Functional matrix-targeting sequences may be as short as nine or twelve amino acids [6,7] and have also been identified as part of the presequence of an intermembrane space protein or as part of the amino terminal sequence of an outer membrane protein (see below). Matrix-targeting sequences transport proteins specifically to the mitochondrial matrix [4].

Conservation of specific primary amino acid sequences is not required for a functional matrix-targeting sequence. Allison and Schatz [8] showed that functional matrix-targeting sequences can be constructed by using only three different amino acids: arginine (a basic amino acid), leucine (an apolar amino acid), and serine (a hydroxylated amino acid). These artificial presequences, like most natural presequences, could form positively charged amphiphilic structures [3,8,9]. As shown by

TABLE I. The Functional Presequences Derived From the Mouse DHFR Gene Resemble Natural Presequences More Closely Than Random Sequences Derived From DHFR

	DHFR Functional		DHFR Random				
	Inserts	Presequences	Inserts	Presequences	Authentic DHFR	Average Proteins	Natural Presequences
% of basic amino acids	11.4	13.5	14.5	15.2	13.4	11.1	17.9
% of acidic amino acids	2.5	1.6	8.0	5.0	8.1	11.1	0.0
Range of hydrophobicity	—	−0.23→ +0.34	—	−0.52→ +0.57	—	—	−0.25→ +0.06
% with $\mu max \geqslant 0.5$	—	80	—	40	—	—	90

Short DNA fragments were generated by DNAse I digestion of the DNA encoding mouse DHFR and fused to the 5′-end of a gene encoding a truncated precursor of subunit IV of cytochrome *c* oxidase. The truncated subunit IV precursor lacks most of its presequence and by itself remains in the cytosol. The inserted DNA sequences used in the calculation encoded short peptides that, possibly together with the remainder of the authentic subunit IV presequence, formed functional matrix-targeting sequences (DHFR functional). The characteristics of such experimentally derived composite presequences were compared to a set of 30 theoretically derived random sequences (DHFR random). These random sequences were generated by taking short reading frames of comparable length to the selected ones, starting at randomly chosen points in the DHFR gene sequence. As with the selected sequences, the random sequences were derived from all six possible reading frames. The inserted sequences were analyzed by themselves (inserts), ie, only the amino acids derived from the DHFR gene, and as composite amino acid sequences including the seven amino acids from the residual presequence of subunit IV and two amino acids created at the fusion points of each construct by the cloning procedure (presequences) [see also 11]. Authentic DHFR, the calculated data for all possible open reading frames encoded in the DNA sequence; Average proteins, the average of a pool of 185 proteins taken from reference [34]. Natural presequences, the combined data for ten natural mitochondrial presequences [11]. For each presequence, μ_{max} is the calculated μ value of the ten-amino-acid segment of maximal hydrophobic moment. This value gives an indication of the potential amphiphilicity of the presequence, assuming an alpha-helical conformation [35]. The value for the hydrophobicity of each presequence was obtained by averaging the hydrophobicity of each segment of ten amino acids obtained by moving a ten-amino-acid window through the presequence [35]. The range of these values is indicated. The narrower the range, the more the fragments resemble each other in their hydrophobicity. Negative values, hydrophilic sequences; positive values, hydrophobic sequences. This table summarizes data presented in reference [11].

Roise et al [9], a natural amphiphilic presequence peptide inserts efficiently into lipid monolayers and bilayers and is able to interact with mitochondrial membranes. The functional artificial presequences are also surface active [8]. Thus, a clear correlation exists between the peptide's amphiphilic character and its function as a mitochondrial matrix-targeting sequence. A positively charged amphiphilic structure may therefore be an important characteristic of a functional matrix-targeting sequence.

MANY SEQUENCES IN EUKARYOTIC AND PROKARYOTIC DNA POTENTIALLY ENCODE MITOCHONDRIAL MATRIX-TARGETING SEQUENCES; THEY CAN BE ACTIVATED BY TRANSPOSITION IN FRONT OF A PROTEIN-CODING SEQUENCE

What is the evolutionary origin of presequences? According to the endosymbiont theory [reviewed in 10], mitochondria and chloroplasts developed from free-living bacteria that entered into symbiosis with other organisms. With time, many of the genes originally present in the bacterial genome were transferred to the nucleus, and the former bacterial genome evolved into the present-day mitochondrial or chloroplast genomes. After gene transfer, the nuclear-encoded proteins had to be reimported into the organelles from the cytosol and, therefore, had to acquire presequences.

Since presequences share only general characteristics, it was not unreasonable to expect that moderately complex genomes such as those of a protoeukaryotic cell or a present-day bacterium might encode many sequences that could function as mitochondrial presequences if placed at the amino terminus of a protein. This idea was indeed correct. Random DNA fragments of about 60–100 bases long were generated from the *Escherichia coli* genome or from the coding region for the mouse cytosolic enzyme dihydrofolate reductase (DHFR). These fragments were fused to the 5′-end of the gene encoding a truncated precursor of subunit IV of cytochrome *c* oxidase. This truncated subunit IV precursor lacked most of its presequence and, as a result, remained in the cytosol. Many DNA fragments were found that restored a functional presequence to the truncated subunit IV precursor. These presequences transported subunit IV into the yeast mitochondrial matrix [11]. Table I summarizes the characteristics of 20 presequences obtained this way from the DNA-encoding mouse DHFR. Most of the reading frames that restored a functional presequence were derived from DNA sequences that are normally not expressed in their natural environment [11]. As with many naturally occurring presequences, these presequences were low in acidic amino acids and neither extremely hydrophobic nor hydrophilic. All contained at least two positively charged residues. Many of these sequences were also predicted to be able to form an amphiphilic alpha-helix (Table I). Clearly, these sequences had general characteristics also found in naturally-occurring presequences. The primary sequences of these presequences are published elsewhere [11]. The functional presequences derived from the *E. coli* DNA had characteristics very similar to those listed in Table I [11].

X-ray crystallography data show that the mouse cytosolic enzyme DHFR contains an amphiphilic alpha-helix whose apolar face is not exposed to the outside of the folded molecule [12]. By gene-fusion experiments, Hurt and Schatz [12] placed this sequence at the amino terminus of either DHFR or of mature cytochrome oxidase subunit IV and showed that the resulting fusion proteins were specifically transported into the mitochondrial matrix in living yeast cells [12]. Thus an amphiphilic helix within a folded protein can be made to function as a mitochondrial matrix-targeting presequence. We believe that many such sequences exist in eukaryotic and prokaryotic genomes, but that most of these sequences are not expressed, or if expressed are buried within the folded structure of proteins. However, such sequences represent a large pool of potential targeting sequences that can be activated by transposition and then refined by natural selection.

THE PRESEQUENCE OF A CHLOROPLAST PROTEIN FUNCTIONS AS A MATRIX-TARGETING SEQUENCE IN YEAST

The presequence of the small subunit of ribulose-1,5-bisphosphate carboxylase/oxygenase (Rubisco, a chloroplast protein) is surprisingly similar to mitochondrial matrix-targeting sequences [13]. In plants, chloroplasts and mitochondria are present in the same cell and each of these organelles has a specific set of proteins. How similar are import sequences of chloroplast and mitochondrial precursor proteins and how are proteins sorted between these organelles? Hurt et al [13] fused part of the presequence of Rubisco to DHFR or to the mature subunit IV of cytochrome *c* oxidase. In yeast cells, the chloroplast presequence fragment transported both proteins into the mitochondrial matrix. Hurt et al [14] also showed that the mature part of the small subunit of Rubisco can be transported into the mitochondrial matrix by a mitochondrial presequence: it lacks a signal that would interfere with its transport into yeast mitochondria. Plants must have a system to distinguish between chloroplast and mitochondrial precursor proteins. How this distinction is made, however, is not known.

MITOCHONDRIAL PRESEQUENCES CONTAIN SEPARATE DOMAINS FOR INTRACELLULAR TARGETING AND FOR INTRAMITOCHONDRIAL SORTING

The presequence of cytochrome c_1 (an inner membrane protein exposed to the intermembrane space) and the amino terminus of the 70-kd outer membrane protein (a protein anchored in the mitochondrial outer membrane and exposed to the cytosol) contain potential matrix-targeting sequences at their amino termini: the first 12 amino acids of the 70-kd protein and the first 16 amino acids of the cytochrome c_1 presequence transport the mouse cytosolic enzyme DHFR into the mitochondrial matrix [15–17]. The entire cytochrome c_1 presequence (61 amino acids), however, transports DHFR and cytochrome *c* oxidase subunit IV, lacking its own presequence (mature subunit IV), into the intermembrane space [16–18], while the first 41 amino acids of the 70-kd protein anchor *E. coli* beta-galactosidase to the mitochondrial outer membrane [19]. Both transport sequences contain, in addition to a matrix-targeting sequence, a sequence responsible for intramitochondrial sorting [see also 4]. These sorting sequences each include a stretch of uncharged amino acids, which is 19 amino acids long for the cytochrome c_1 presequence [20] and 28 amino acids for the 70-kd protein [21]. Each of these stretches is long enough to span a lipid bilayer if an alpha-helical structure is assumed.

Deletions have been made in the cytochrome c_1 presequence to define the matrix-targeting and intramitochondrial sorting domains more precisely [17]. The results are summarized in Figure 1. The first 16 amino acids (the first quarter) of the

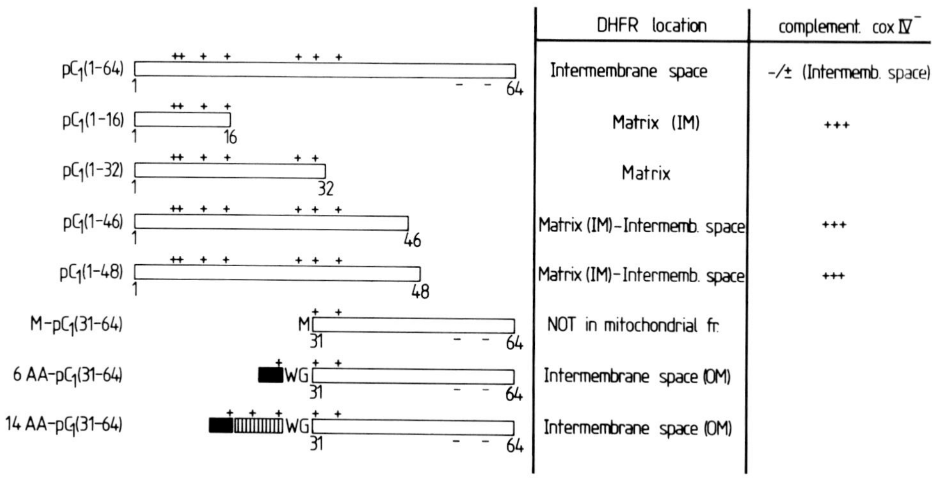

Fig. 1. The cytochrome c_1 presequence contains separate matrix-targeting and intramitochondrial sorting domains. By gene manipulation, parts of the cytochrome c_1 presequence (open boxes) were deleted or replaced by other sequences to determine the functions of these parts in intracellular and intramitochondrial sorting. Each of the sequences shown in the left panel was fused to the amino terminus of mouse DHFR. Some were also fused to the amino terminus of a cytochrome oxidase subunit IV protein lacking most of its own presequence. The intracellular and intramitochondrial locations of the fusion proteins were assayed in yeast cells expressing these proteins and/or after protein import into isolated mitochondria. The fusions to subunit IV were also tested for their ability to complement the growth deficiency of cells lacking the gene for the authentic subunit IV precursor (cox IV$^-$ cells). Thus far, all subunit IV proteins transported into the matrix complement the mutant's inability to grow on glycerol. The positions of basic (+) or acidic (−) amino acids are indicated. Filled box, amino acid sequence M-L-S-R; hatched box, amino acid sequence L-S-L-R-L-L-S-R. 6AA and 14AA, artificial matrix-targeting sequences containing 6 and 14 amino acids, respectively. Matrix (IM), protein bound to inner membrane but exposed to the matrix; intermembrane space (OM), mature-sized DHFR in intermembrane space and uncleaved precursor at the outer surface of the outer membrane. This figure summarizes data presented in references [16] and [17].

presequence transport attached DHFR or mature subunit IV efficiently to mitochondria and into the matrix: it is a matrix-targeting sequence. The second half of the cytochrome c_1 presequence by itself (starting at amino acid 31) does not contain sufficient information for transporting an attached protein to mitochondria: it lacks information for intracellular targeting. The last quarter of the presequence (starting at amino acid 49) is essential for intramitochondrial sorting. Our data suggest that the entire stretch of uncharged amino acids (amino acids 36–54) may be part of the intramitochondrial sorting domain. The matrix-targeting and intramitochondrial sorting functions are, thus, located in different parts of the cytochrome c_1 presequence.

The first half of the cytochrome c_1 presequence (30 amino acids) can be replaced by artificial matrix-targeting sequences as short as six amino acids without abolishing transport of attached DHFR into the intermembrane space (Fig. 1). This shows that the first 30 amino acids of the authentic cytochrome c_1 presequence do not contain a hidden sequence motif which acts together with the second half of the presequence in intramitochondrial sorting.

THE INTRAMITOCHONDRIAL SORTING SEQUENCE IN THE CYTOCHROME c_1 PRESEQUENCE IS A STOP-TRANSPORT SEQUENCE SPECIFIC FOR THE INNER MEMBRANE

How does the intramitochondrial sorting sequence in the cytochrome c_1 presequence work? Is it a stop-transport sequence specific for the inner membrane as suggested [4,20]? If so, then the attached protein should cross the mitochondrial outer, but not the inner, membrane. Alternatively, the precursor protein may first be completely transported into the mitochondrial matrix by the matrix-targeting sequence and subsequently be reexported into the intermembrane space. This latter model was recently invoked [22] for the iron-sulphur protein precursor of the *Neurospora crassa* cytochrome bc_1 complex. This protein, like cytochrome c_1, is assembled into the transmembrane cytochrome bc_1 complex in such a way that it is partly exposed toward the intermembrane space [22].

To test which pathway is followed by the cytochrome c_1 presequence, we compared the import of (1) a fusion protein between the cytochrome c_1 presequence and mature subunit IV of cytochrome *c* oxidase (pC_1 (1–64)-COX IV), (2) the authentic subunit IV precursor, and (3) the authentic yeast iron-sulphur protein precursor, into both mitochondria and mitoplasts (mitochondria in which the outer membrane has been disrupted). Both the authentic subunit IV precursor (not shown) and iron-sulphur protein precursor (Fig. 2, bottom panel) and their cleavage intermediates became protease resistant in mitochondrial and mitoplast import: they were transported across the inner membrane. In contrast, the pC_1 (1–64)-COX IV precursor and its cleavage intermediate became protease resistant upon import into mitochondria but not upon import into mitoplasts (Fig. 2, top panel), although mitoplasts generated as much of the intermediate as mitochondria did [18]. Thus, the iron-sulphur precursor and the pC_1 (1–64)-COX IV precursor use different pathways to reach the intermembrane space. The intramitochondrial sorting sequence in the cytochrome c_1 presequence indeed behaves as a stop-transfer sequence specific for the inner membrane.

TRANSPORT OF PROTEINS INTO MITOCHONDRIA REQUIRES ATP AS WELL AS AN ENERGIZED INNER MEMBRANE

Transport into chloroplasts and the endoplasmic reticulum, and protein export from bacteria, requires ATP. Is ATP also required for protein import into mitochondria? To test this, a fusion protein consisting of the first 22 amino acids of the subunit IV presequence fused to DHFR was purified to near homogeneity (pCOX IV(1–22)-DHFR) [23]. The availability of a radiochemically pure precursor protein allowed the study of protein import in the absence of a cell-free protein synthesizing system and, thus, of ATP. Efficient import into isolated mitochondria was obtained if the respiratory substrates succinate and malate were used for generating the membrane potential and for ATP synthesis (Fig. 3, lanes 2 and 3). If ATP synthesis and transport of ATP into the matrix were specifically inhibited with oligomycin (an inhibitor of the mitochondrial ATPase) and carboxyatractyloside (an inhibitor of adenine nucleotide translocation across the mitochondrial inner membrane), no import was observed (Fig. 3, lanes 6 and 7). Addition of ATP (Fig. 3, lanes 8 and 9) or GTP, but not of

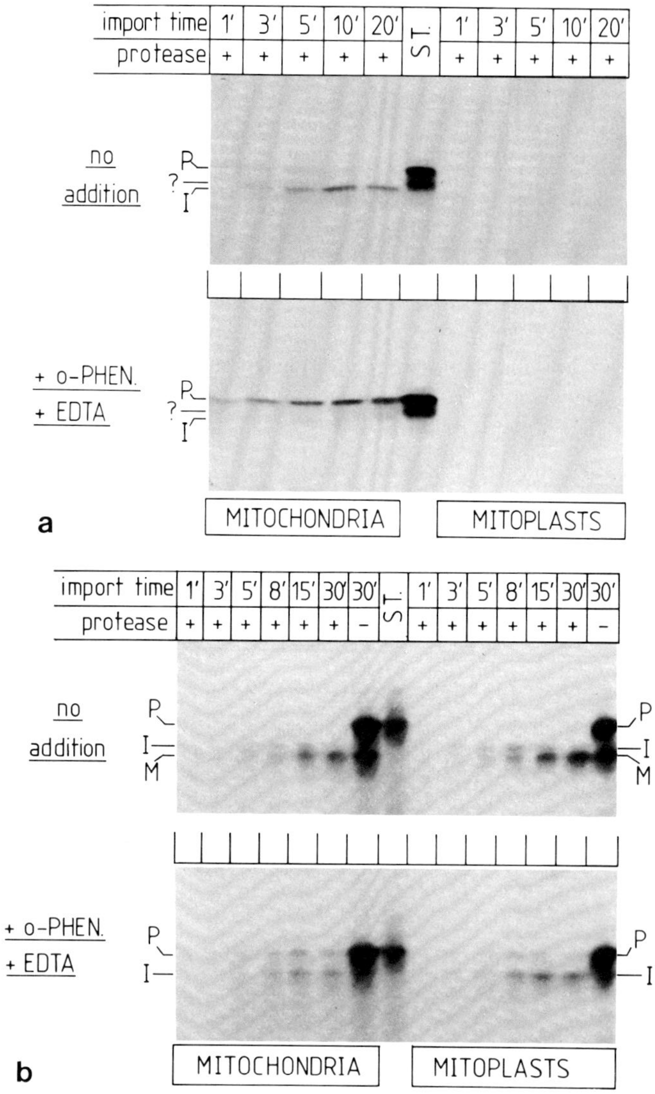

Fig. 2. The intramitochondrial sorting sequence in the cytochrome c_1 presequence is a stop-transport sequence specific for the inner membrane. The radiolabeled precursor of the iron-sulphur protein of the cytochrome bc_1 complex (bottom panel) or the $pC_1(1-64)$-COX IV fusion protein (top panel) were incubated with energized mitochondria or mitoplasts (mitochondria whose outer membranes have been broken) either in the absence (no addition) or the presence of ortho-phenanthroline plus EDTA (+o-PHEN. +EDTA) to inhibit the matrix-located processing protease. At the indicated times, samples were taken and diluted into ice-cold buffer containing various inhibitors that interfere with the functions of the respiratory chain, the mitochondrial ATPase, and the matrix-located processing protease. Where indicated, the buffer also contained trypsin to remove all proteins bound to the mitochondrial surface or outside the inner membrane in mitoplasts. After incubation, the organelles were reisolated and analyzed for the presence of radiolabeled proteins by SDS-polyacrylamide gel electrophoresis and fluorography. ST., 20% of the radiolabled precursor added to each of the other incubations; P, precursor; I, cleavage intermediate; M, mature-size iron-sulphur protein; ?, an unknown radiolabeled contaminant present in this particular preparation of the $pC_1(1-64)$-COX IV precursor. This contaminant does not bind to mitochondria and its electrophoretic mobility is distinctly lower than that of the cleavage intermediate. This figure summarizes data presented in reference [18].

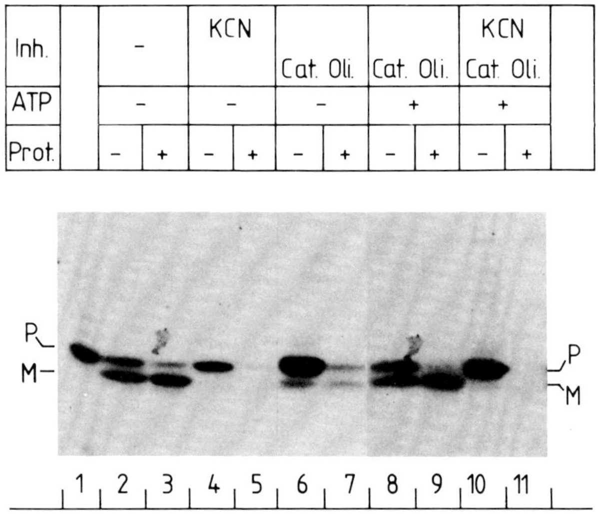

Fig. 3. Import of a purified precursor protein into mitochondria requires both a membrane potential and ATP. Purified radiolabeled pCOX IV(1–22)-DHFR fusion protein was incubated with isolated yeast mitochondria in the presence of the respiratory substrates succinate and malate, and, where indicated, with ATP and/or the inhibitors (inh.) mentioned below. Mitochondria were either left untreated (lanes 2, 4, 6, 8, 10) or treated with proteinase K (Prot.) after import (lanes 3, 5, 7, 9, 11), to digest all proteins bound to the mitochondrial surface. The mitochondria were then reisolated by centrifugation and analyzed for the presence of radiolabeled proteins by SDS-polyacrylamide gel electrophoresis and fluorography. Lane 1 contains a standard corresponding to 10% of the amount of precursor added to each of the other incubations. Cat., carboxyatractyloside (an inhibitor of adenine nucleotide translocation across the mitochondrial inner membrane); Oli., oligomycin (an inhibitor of the mitochondrial ATPase); P, precursor; M, mature-size DHFR protein. This figure summarizes data presented in reference [24].

TTP, CTP, or a noncleavable ATP analogue, restored import [24]. The restoration of import by ATP could not be explained by an effect on the membrane potential: ATP also supported import if it was added in the presence of oligomycin and carboxyatractyloside (Fig. 3, lanes 8 and 9). In the absence of a membrane potential, ATP did not support protein import: inhibition of respiration by cyanide blocked import in the absence and in the presence of ATP (Fig. 3, lanes 4, 5, 10, and 11). Similar results were presented recently by Pfanner and Neupert [25] and Chen and Douglas [26]. Differences were only observed in the nucleotides that could replace ATP. These results all demonstrate that, in addition to the membrane potential, ATP is required for transporting proteins across mitochondrial membranes.

TRANSPORT OF PROTEINS ACROSS MITOCHONDRIAL MEMBRANES CAN BE INHIBITED BY STABILIZING THE PROTEIN'S STRUCTURE

Many precursor proteins can be imported into mitochondria post-translationally [reviewed in 1]. What is the structure of the precursor protein prior to its transport into mitochondria? The purified pCOX IV(1–22)-DHFR protein is enzymatically active and has K_m values very similar to that of authentic DHFR without a presequence [23]. This purified protein is efficiently imported by isolated mitochondria. Mouse DHFR found in yeast mitochondria subsequent to import is enzymatically active [7,15–17]. The crystal structure of the highly homologous chicken DHFR shows that the enzyme has a very stable globular structure. Is this structure maintained during transport across the membranes? To answer this question, import of various DHFR-containing precursors into isolated mitochondria was studied. Import of each of these precursors could be inhibited by methotrexate (Fig. 4, top panel shows this for a fusion protein between the first 27 amino acids of the ADHIII precursor and DHFR). This inhibitor binds very tightly to the active site of DHFR and "freezes" the folded conformation of the protein. With the purified pCOX IV(1–22)-DHFR protein, inhibition of import and DHFR enzyme activity was observed with similar methotrexate concentrations (not shown). The inhibition was specific for DHFR-containing fusion proteins and was not observed for several other imported proteins (Fig. 4, bottom panel shows this for the authentic pADHIII precursor). The inhibition

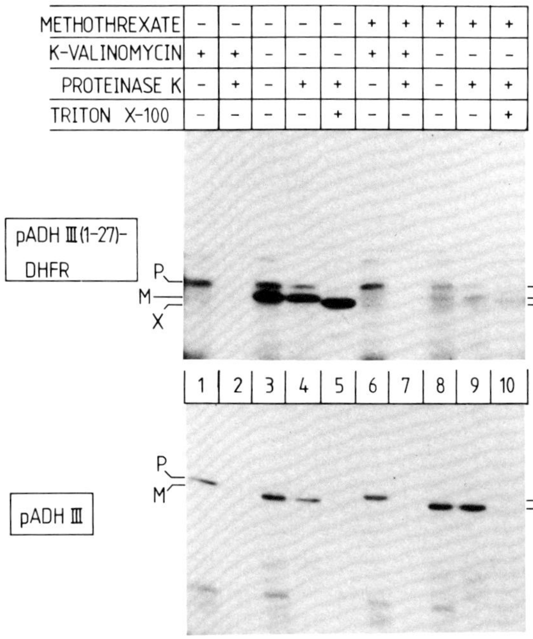

Fig. 4. Methotrexate inhibits the import into isolated mitochondria of the pADHIII(1–27)-DHFR fusion protein but not that of the authentic pADHIII precursor. The pADHIII(1–27)-DHFR fusion protein (top panel) and the authentic pADHIII precursor (bottom panel) were synthesized by transcription-translation in the presence of ^{35}S-methionine. The radiolabeled precursors were incubated with energized mitochondria (lanes 3–5 and 8–10) or with mitochondria deenergized with valinomycin and KCl (lanes 1, 2, 6, 7). The incubation was done in the absence (lanes 1–5) or presence (lanes 6–10) of methotrexate, an inhibitor of DHFR which binds tightly to the enzymatically active protein. After incubation, some samples were treated with proteinase K in the absence (lanes 2, 4, 7, 9,) or presence of the detergent Triton X-100 (lanes 5, 10). The mitochondria were then reisolated by centrifugation and tested for the presence of radiolabeled proteins by SDS-polyacrylamide gel electrophoresis and fluorography. P, precursor; M, mature-size protein. Band X is a proteinase K–resistant degradation product of DHFR which is of distinctly lower molecular weight than imported "mature-sized" DHFR. It is not generated upon protase treatment of intact mitochondria but only upon protease treatment of detergent-lysed mitochondria [16–18, 23, 24, 31]. Band X is also generated if a purified, DHFR-containing precursor protein is treated at low protease concentrations in the absence of mitochondria. Further digestion of band X then requires high protease concentrations (T. Endo and S. Hwang, unpublished). This figure summarizes data presented in reference [23].

of import is not the result of the inhibitor's inability to be transported across membranes. An apolar inhibitor, which by itself can freely cross membranes, also inhibited import of DHFR-containing precursor molecules [23]. Thus, the protein may have to undergo a structural change and at least partially unfold in order to move across the mitochondrial membranes.

ATP MAY BE NEEDED TO CHANGE PRECURSOR STRUCTURE DURING MEMBRANE TRANSPORT

Is there a correlation between the requirement for ATP cleavage and the change in the protein's structure during transport across the membrane? It is tempting to speculate that ATP may be needed for changing the compact DHFR molecule into a transport-competent structure. To test this hypothesis, the pCOX IV(1–22)-DHFR protein was synthesized in a cell-free protein-synthesizing system and cycloheximide was added after a few minutes of protein synthesis to generate a collection of growing peptide chains. These polypeptides were shown to be nascent chains of the pCOX IV(1–22)-DHFR fusion protein [27]. The nascent chains were separated from the cell-free system (which contains high concentrations of ATP to support protein synthesis) by pelleting the ribosomes, washing them once in the presence of an ATP trap, and then releasing the nascent chains by dissociation of the ribosomes with EDTA. The ATP-depleted nascent chains were then mixed with the purified pCOX IV(1–22)-DHFR precursor, and import of the ATP-depleted protein mixture into isolated mitochondria was studied in the presence or absence of added ATP [27]. Figure 5 shows that import of the full-length precursor is highly ATP-dependent: imported, mature-size protein is formed only if ATP was present during protein import. In contrast, most but not all of the lower-molecular-weight polypeptides were still imported into mitochondria if no ATP was added. The import efficiency of nascent chains, irrespective of whether ATP was added or not, was much higher than that of the full-length precursor protein: for the precursor protein 5–10% but for the nascent chain polypeptides much more than 20% of the total amount added to mitochondria was imported. Thus, import of nascent precursor chains into mitochondria is highly efficient and only slightly, if at all, ATP-dependent. Additional experiments [27] showed that the nascent chains, in contrast to the completed protein, were highly sensitive to protease in the absence of mitochondria. This suggests that they did not have the compact, protease-resistant structure characteristic of the completed pCOX IV(1–22)-DHFR precursor. These experiments demonstrate a correlation between lack of stable structure and lower ATP requirement for protein import into isolated mitochondria, suggesting that ATP is needed for protein unfolding.

PROTEIN IMPORT INTO ISOLATED MITOCHONDRIA IS INHIBITED BY ANTIBODIES RAISED AGAINST A SPECIFIC SUBSET OF OUTER MEMBRANE PROTEINS

Are accessory proteins required for transport of precursors into mitochondria? The existence of such proteins has been inferred from observations that treatment of the mitochondrial surface with different proteases abolished protein import without disrupting the outer membrane [28–30]. This approach, however, is not suited for isolation proteins of a putative translocation machinery. Ohba and Schatz [31] recently showed that the combination of a mild protease pretreatment of mitochondria and

ATP	+	+	–	–	+	–
K-VALINOMYCIN	–	–	–	–	+	+
PROTEINASE K	+	+	+	+	+	+
TRITON X-100	–	+	–	+	–	–

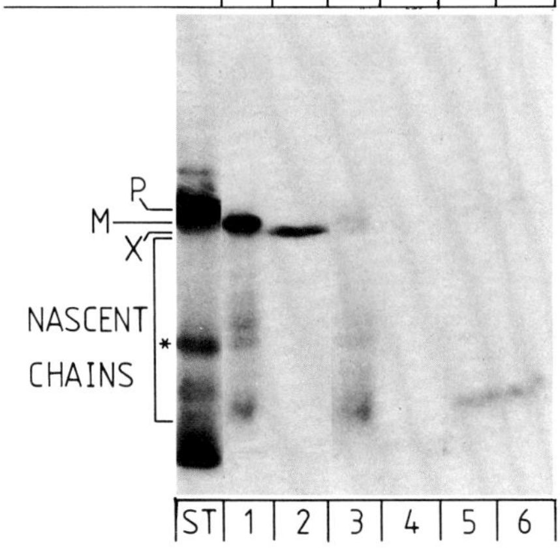

Fig. 5. Only completed, but not nascent chains of the pCOX IV(1–22)-DHFR fusion protein require added ATP for import into isolated yeast mitochondria. Radiolabeled nascent chains of the pCOX IV(1–22)-DHFR precursor were isolated and mixed with the radiolabeled, purified, completed pCOX IV(1–22)-DHFR precursor. The mixture was incubated with energized yeast mitochondria (lanes 1–4) or mitochondria deenergized with valinomycin and KCl (lanes 5, 6). All incubations were done in a shaking waterbath in the presence of the respiratory substrate succinate to generate a membrane potential. Efrapeptin (an inhibitor of the mitochondrial ATPase) was added to each incubation to inhibit formation of ATP by oxidative phosphorylation. Mitochondria were then incubated with proteinase K to remove all proteins bound to the mitochondrial surface (lanes 1, 3, 5, 6). Protease treatment was sometimes done in the presence of Triton X-100 (lanes 2, 4). Each sample was analyzed by SDS polyacrylamide gel electrophoresis and fluorography. ST, 20% of the amount of nascent chains and completed presursor protein added to each of the other incubations; P, precursor; M, mature-size protein; X, proteinase K–resistant digestion product of DHFR (see legend to Fig. 4); band marked with (*), radiolabeled contaminant present in this preparation of the purified full-length precursor protein. The nature of the lower-molecular-weight band present in lanes 5 and 6 is not known. This figure summarizes data presented in reference [27].

incubation with Fab fragments of an antiserum directed against the 45-kd proteins of the mitochondrial outer membrane virtually abolished import of the purified pCOX IV(1–22)-DHFR protein into isolated mitochondria. This import block was not observed with Fab fragments of antibodies directed against other outer membrane proteins. The import block was not specific for the pCOX IV(1–22)-DHFR fusion protein; it was also observed for the pADHIII(1–27)-DHFR precursor or the authentic subunit IV precursor. The inhibitory effects of anti-45-kd antibodies was abolished by preincubating them with purified yeast outer membranes. These experiments suggest that a 45-kd protein of the yeast mitochondrial outer membrane participates in the translocation of proteins into mitochondria. This observation now makes it possible to start a detailed study of the putative protein receptor(s) and translocation machinery. The machinery responsible for the transmembrane transport still is the major "black box" in the study of protein translocation into mitochondria. It seems reasonable to expect that our knowledge of the machinery will develop rapidly in the next few years, especially since several yeast mutants have been isolated [32,33] which are (partially) blocked in protein transport into mitochondria.

ACKNOWLEDGMENTS

We thank all the members of the Schatz laboratory for their contributions to the pleasant and productive working environment and Prof. Schatz for discussions and critical reading of the manuscript. This work was supported by grants 3.660-0.84 from the Swiss National Science Foundation and CBY-1 1 R01 GM37803-01 from the US Public Health Service to G. Schatz and a long-term EMBO fellowship to A. Baker.

REFERENCES

1. Hay R, Boehni P, Gasser S: Biochim Biophys Acta 779:65, 1984.
2. Douglas M, Takeda M: Trends Biochem Sci 10:192, 1985.
3. Von Heijne G: EMBO J 5:1335, 1986.
4. Hurt EC, van Loon APGM: Trends Biochem Sci 11:204, 1986.
5. Van Loon APGM, Young ET: EMBO J 5:161, 1985.
6. Keng T, Alani E, Guarente L: Mol Cell Biol 6:355, 1986.
7. Hurt EC, Pesold-Hurt B, Suda K, Oppliger W, Schatz G: EMBO J 4:2061, 1985.
8. Allison DS, Schatz G: Proc Natl Acad Sci USA 83:9011, 1986.
9. Roise D, Horvath SJ, Tomich JM, Richards JH, Schatz G: EMBO J 5:1327, 1986.
10. Gray MW, Doolittle WF: Microbiol Rev 46:1, 1982.
11. Baker A, Schatz G: Proc Natl Acad Sci USA 84:3117, 1987.
12. Hurt EC, Schatz G: Nature 325:499, 1987.
13. Hurt EC, Soltanifar N, Goldschmidt-Clermont M, Rochaix J-D, Schatz G: EMBO J 5:1343, 1986.
14. Hurt EC, Goldschmidt-Clermont M, Pesold-Hurt B, Rochaix J-D, Schatz G: J Biol Chem 261:11440, 1986.
15. Hurt EC, Mueller U, Schatz G: EMBO J 4:3509, 1985.
16. Van Loon APGM, Braendli AW, Schatz G: Cell 44:801, 1986.
17. Van Loon APGM, Braendli AW, Pesold-Hurt B, Blank D, Schatz G: EMBO J 6:2433, 1987.
18. Van Loon APGM, Schatz G: EMBO J 6:2441, 1987.
19. Hase T, Mueller U, Riezman H, Schatz G: EMBO J 3:3157, 1984.
20. Sadler I, Suda K, Schatz G, Kaudewitz F, Haid A: EMBO J 3:2137, 1984.
21. Hase T, Riezman H, Suda K, Schatz G: EMBO J 2:2169, 1983.
22. Hartl F-U, Schmidt B, Wachter E, Weiss H, Neupert W: Cell 47:939, 1986.
23. Eilers M, Schatz G: Nature 322:228, 1986.
24. Eilers M, Oppliger W, Schatz G: EMBO J 6:1073, 1987.
25. Pfanner N, Neupert W: FEBS Lett 209:152, 1986.
26. Chen W-J, Douglas MG: Cell 49:651, 1987.
27. Verner K, Schatz G: EMBO J 6:2449, 1987.
28. Zwizinski C, Schleyer M, Neupert W: J Biol Chem 258:4071, 1983.
29. Riezman H, Hay R, Gasser S, Daum G, Schneider G, Witte C, Schatz G: EMBO J 2:1105, 1983.
30. Zwizinski C, Schleyer M, Neupert W: J Biol Chem 259:7850, 1984.
31. Ohba M, Schatz G: EMBO J 6:2109, 1987.
32. Yaffe MP, Schatz G: Proc Natl Acad Sci USA 81:4819, 1984.
33. Yaffe MP, Ohta S, Schatz G: EMBO J 4:2069, 1985.
34. Dayhoff MO, Dayhoff RE, Hunt LT: In Dayhoff MO (ed): "Atlas of Protein Sequence and Structure." Vol 5, Supplement 2. Silver Spring, MD: National Biomedical Research Foundation, 1976, p 301.
35. Eisenberg D, Schwarz M, Komaromy M, Wall R: J Mol Biol 179:125, 1984.

Journal of Cellular Biochemistry 36:73–81 (1988)
Molecular Biology of Intracellular Protein Sorting and Organelle Assembly 47–55

Inhibition of Endocytosis From Coated Pits by Acidification of the Cytosol

Kirsten Sandvig, Sjur Olsnes, Ole W. Petersen, and Bo van Deurs

Institute for Cancer Research at The Norwegian Radium Hospital and The Norwegian Cancer Society, Montebello, 0310 Olso, Norway (K.S., S.O.), Department of Anatomy, The Panum Institute, University of Copenhagen, DK-2200 Copenhagen N, Denmark (O.W.P., B.v.D.)

Binding and endocytosis of the ligands transferrin, epidermal growth factor (EGF), and ricin were measured in a number of different cell lines after treatment of cells with compounds that react with SH-groups and under conditions where the cytosolic pH was lowered. N-ethylmalemide and diamide irreversibly inhibited endocytosis of all ligands tested, whereas low pH in the cytosol strongly inhibited endocytosis of transferrin and EGF. Data obtained by electron microscopy indicated that the formation of coated vesicles from coated pits is inhibited in acidified cells. Entry of ricin was much less affected, and ricin endocytosed under these conditions was able to intoxicate the cells. At low pH in the cytosol there was a calcium-dependent increase in the number of transferrin receptors at the cell surface. The increase was even larger in the presence of the calcium ionophore A23187, whereas it was completely blocked by the calmodulin antagonists trifluoperazine and W7. The results show that endocytosis from coated pits can be inhibited in a reversible way by acidification of the cytosol and they suggest that a second pathway of endocytosis exists, possibly involving formation of vesicles from uncoated areas of the membrane.

Key words: endocytosis, coated pits, cytosol, transferrin, epidermal growth factor

During recent years there has been an ongoing discussion as to whether endocytosis occurs exclusively by the coated pit/coated vesicle pathway or whether surface-bound ligands and fluid phase markers can enter cells by formation of vesicles from uncoated areas of the membrane. Results obtained by several investigators suggest that endocytosis of surface-bound ligands occurs not only from coated areas, but also from noncoated areas of the membrane. Thus, it was recently shown by Moya et al [1] that endocytosis of the toxic lectin ricin continued when the coated pits had been removed by potassium depletion of the cells. Also, infection of cells with human rhinovirus type 2, which requires passage through a low pH compartment for entry into the cytosol, occurs normally in potassium-depleted cells [2], suggesting

Received March 6, 1987; revised and accepted July 22, 1987.

that transfer of this virus to a low pH compartment occurs even in the absence of coated pits. Furthermore, Morris et al [3,4] have found that pseudomonas toxin and diphtheria toxin are internalized even when endocytosis of the toxins from coated pits is inhibited by addition of amines, and it has been suggested that smooth pits are involved in the uptake of cholera toxin and tetanus toxin [5].

Removal of coated pits by potassium depletion of the cytosol has a number of other effects on the cells, like inhibition of chloride transport, removal of the membrane potential, and cell shrinkage [6]. Alternative methods to inhibit endocytosis are therefore required. In the present work we have studied the effect of a number of treatments on the rate of endocytosis. For this purpose we have measured the endocytic uptake of transferrin and EGF, which by a number of groups have been shown to be endocytosed from coated pits [7–10], as well as the uptake of the lectin ricin, which binds to terminal galactose on both glycolipids and glycoproteins, and which is therefore likely to be endocytosed by all different mechanisms operating (for review, see [11]).

MATERIALS AND METHODS

Materials

[^{14}C]Dimethyloxazolidine 2,4-dione (DMO) was obtained from New England Nuclear (Boston, MA). [^{3}H]Leucine, [^{14}C]sucrose, [^{3}H]H_2O, and Na ^{125}I were from the Radiochemical Centre (Amersham, UK). Transferrin, pronase, epidermal growth factor, 2-(N-morpholino) ethane sulfonic acid (MES), Hepes, Tris, trifluoperazine, nigericin, and valinomycin were obtained from Sigma Chemical Co. (St. Louis, MO). W7 and W5 were purchased from Seikagaku Kogyo Co., LTD (Tokyo, Japan). Amiloride was a gift from Merck, Sharp & Dohme (Drammen, Norway). Transferrin was saturated with iron as described [12], and ^{125}I-labeled ligands were prepared by the iodogen method [13].

Conjugates of ricin and colloidal gold (particle size 5 nm) were prepared by the method of Slot and Geuze [14] and the amount of protein necessary to stabilize the colloidal gold solution was determined by the method of Horisberger and Rosset [15].

Cells

Hep-2 cells (human laryngeal carcinoma cells with HeLa markers) were obtained from Dr. P. Boquet (Institute Pasteur, Paris, France). A431 cells were obtained from the American Tissue Type Collection (Rockville, MD). HeLA S3, Vero, and MCF-7 cells are strains that have been growing in this laboratory for years. All cell lines used were maintained as monolayer cultures in minimal essential medium complemented with penicillin, streptomycin, and 10% (v/v) fetal calf serum and in an atmosphere containing 5% CO_2. The day before use the cells were seeded out into 24-well disposable trays, Petri dishes, or into T-25 flasks as indicated in legends to figures.

Measurement of Receptor-mediated Endocytosis of ^{125}I-Transferrin, ^{125}I-EGF, and ^{125}I-Ricin

The amount of surface-bound and internalized transferrin was measured as described by Ciechanover et al [16]. Briefly, cells were incubated with ^{125}I-transferrin at 37°C for the indicated periods of time, washed three times with ice-cold PBS, and then treated for 1 hr at 0°C with 0.3 ml serum-free medium containing 0.3% (w/v) pronase. Then the cells and the medium were transferred to Eppendorf tubes and centrifuged for 2 min, and the radioactivity in the pellet and in the supernatant was measured.

Endocytosis of ^{125}I-EGF was measured as described by Haigler et al [17]. For this purpose, cells incubated with ^{125}I-EGF were washed five times with ice-cold PBS, and then incubated for 6 min in a solution containing 0.5 M NaCl, 0.2 M acetic acid, pH 2.5, to release surface-bound EGF. The cells were subsequently washed once in the same buffer, then dissolved, and finally the cell-associated radioactivity was measured.

Endocytosis of ^{125}I-labeled ricin was measured as the amount of toxin that could not be removed from the cells with lactose, as previously described [18].

Measurement of pH in the Cytosol

The cytosolic pH was determined from the distribution of the weak acid [^{14}C]DMO as described by Deutsch et al [19]. Cells were incubated with [^{14}C]DMO (5 μCi/ml) for 5 min at 37°C, then washed five times with ice-cold PBS, and the cell-associated radioactivity was measured. The volume of extracellular liquid associated with the cells after washing five times was measured after incubation of the cells with [^{14}C]sucrose (5 μCi/ml). The extracellular space corresponded to 7% of the cell volume. The total volume of the cells in a well was calculated from the amount of [^{3}H]H_2O associated with cells after incubation with 1 μCi/ml [^{3}H]H_2O for 18 hr.

Electron Microscopical Studies

Cells growing in monolayers in T25 flasks were washed with Hepes buffer and incubated with a ricin-gold conjugate (Ri-Au) in Hepes medium for 30 min at 37°C, as previously described [20,21]. Following incubation, the cultures were washed and fixed with 2.5% glutaraldehyde in 0.1 M Na-cacodylate buffer, pH 7.2, for 60 min at room temperature. The fixed cells were scraped off the flasks and centrifuged in buffer for 25 min at 1,600g. Pellets were postfixed with 2% OsO_4 in cacodylate buffer, pH 7.2, for 1 hr at 4°C and treated with 1% uranyl acetate in distilled water for 1 hr at 20°C. Cell pellets were thereafter embedded in Epon, cut at approximately 50 nm, and examined in a JEOL 100 CX electron microscope.

The frequency of coated pits was determined by counting any coated pit located at the plasma membrane and less than three profile diameters away from the plasma membrane.

RESULTS AND DISCUSSION

It is well established that transferrin binds to surface receptors that cluster in coated pits. Ricin, on the other hand, binds diffusely to the cell surface. In fact, when Vero cells are incubated with a conjugate of ricin-Au, the conjugate can be seen both in coated pits and in uncoated invaginations of the membrane. This is expected because of the wide variety of molecules that carry terminal galactose and which can therefore serve as binding sites for ricin (Fig. 1).

It has been reported earlier that compounds that modify SH-groups have an inhibitory effect on endocytosis in macrophages [22]. We therefore studied the effect of such compounds on the endocytosis of transferrin and ricin in Vero cells. The data

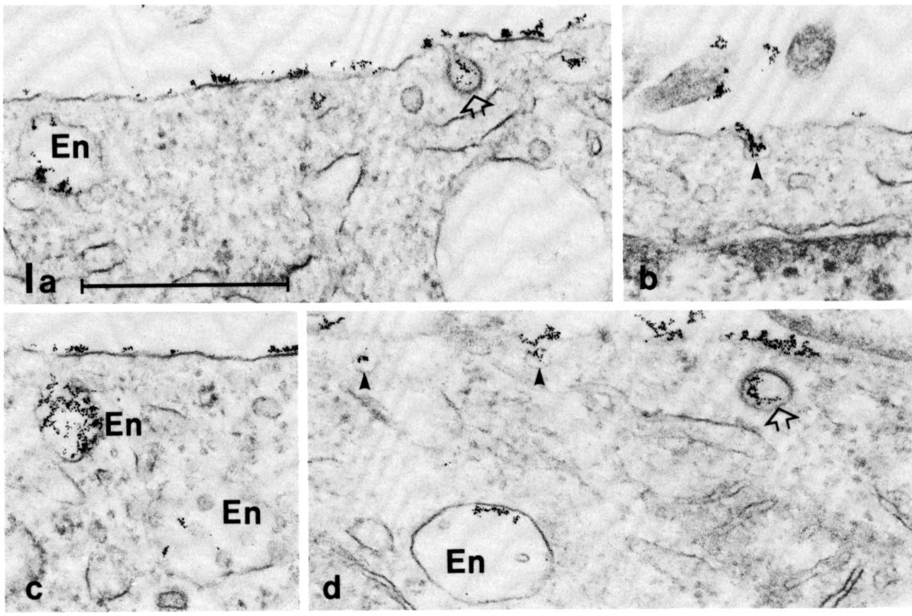

Fig. 1. Surface binding and endocytosis of Ri-Au in Vero cells. Incubation with Ri-Au for 30 min at 37°C. Ri-AU is seen scattered on the cell surface, in coated pits (open arrows) and smooth pits (arrowheads) as well as in endosomes (En). **a–d** ×70,000.

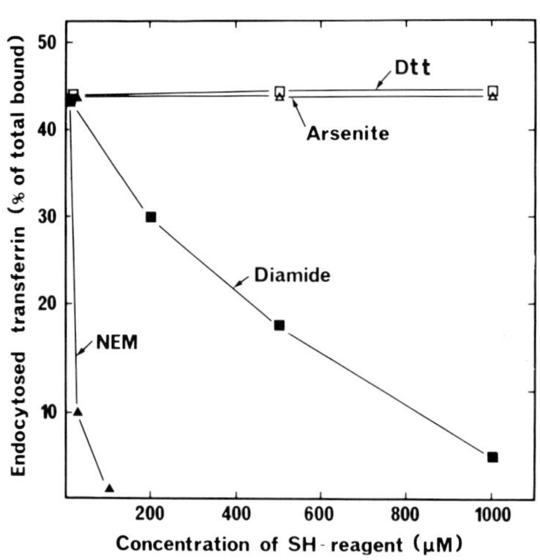

Fig. 2. Ability of SH-reagents to inhibit endocytosis of transferrin. Vero cells growing in 24-well disposable trays were incubated for 10 min at 37°C in Hepes medium (pH 7.2) with and without N-ethylmalemide (NEM), arsenite, diamide, and dithiothreitol (Dtt). ^{125}I-transferrin (200 ng/ml, 38,600 cpm/ng) was then added and, after a 10-min incubation, surface bound and endocytosed transferrin was measured as described in Materials and Methods.

in Figure 2 show that N-ethylmalemide and diamide in submillimolar concentrations strongly inhibited the endocytosis of transferrin. There was no effect of arsenite at similar concentrations, but at a higher concentration (5 mM), arsenite also reduced endocytosis (not demonstrated). On the other hand, dithiotreithol did not have any effect. Also the endocytic uptake of ricin was strongly reduced by N-ethylmalemide and diamide (data not shown) and sulfhydryl reagents are therefore not suitable to study differences in the uptake mechanisms of the two ligands.

We then tested whether or not acidification of the cytosol has different effects on the endocytosis of different ligands. For this purpose, Vero cells, Hep2 cells, and A431 cells were acidified by prepulsing with NH_4Cl. During incubation of the cells with NH_4Cl, presumably both NH_3 and NH_4^+ enter the cytosol from the surrounding medium, but the cells under this incubation are able to regulate the cytosolic pH by using the Na^+/H^+ exchanger. Upon subsequent transfer to medium without NH_4Cl, NH_4^+ in the cytosol will dissociate to NH_3 and H^+, and the membrane permeant NH_3 will then diffuse out of the cells and leave the protons behind [21]. To prevent normalization of the internal pH by the Na^+/H^+ exchanger, the experiment was either carried out in Na^+-free buffer or in the presence of amiloride. As shown in Figure 3, endocytosis of transferrin and EGF was strongly reduced under these conditions.

The extent of acidification obtained was calculated from the distribution of the weak radioactive acid [^{14}C]DMO. We found that preincubation of Vero cells with 10 mM NH_4Cl and subsequent removal of the amine reduced the pH from 6.9 to 6.5, whereas preincubation with 25 mM NH_4Cl reduced the internal pH to approximately 6.0. The data therefore indicate that to reduce the endocytosis significantly, the internal pH must be below pH 6.5. It should be noted that the concentration of NH_4Cl required to give a certain acidification varied to some extent from one day to another.

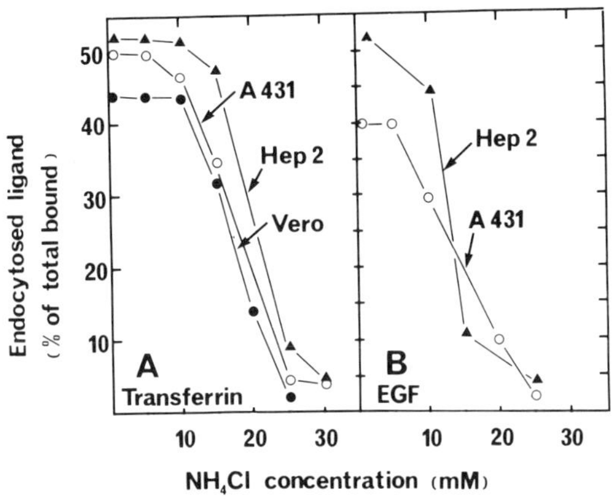

Fig. 3. Effect of acidification of the cytosol by NH_4Cl prepulsing on endocytic uptake of transferrin and EGF. The indicated cell lines growing in 24-well disposable trays **(A)** or in Petri dishes (diameter 35 mm) **(B)** were incubated for 30 min at 37°C in Hepes medium (pH 7.0) with and without the indicated concentrations of NH_4Cl. The medium was then removed, and 0.2 ml **(A)** or 0.5 ml **(B)** 0.14 M KCl containing 2 mM $CaCl_2$, 1 mM $MgCl_2$, 1 mM amiloride, and 20 mM Hepes (pH 7.0) was added. After an additional 5-min incubation, ^{125}I-labeled transferrin (200 ng/ml, 38,600 cpm/ng) **(A)** and EGF (7 ng/ml, 15,000 cpm/ng) **(B)** were added, and endocytosed transferrin and EGF were measured after 10 min or 15 min, respectively.

Electron microscopical studies showed that there was a similar number of coated pits in cells with acidified cytosol and in control cells (not shown). Furthermore, in experiments with a HRP-labeled second antibody against an antitransferrin receptor antibody, it was demonstrated that the coated pits in acidified cells contained transferrin receptors. The data therefore suggest that the reduced endocytosis is due to inhibition of formation of coated vesicles from coated pits.

When we measured endocytosis of ^{125}I-ricin in acidified cells, we found that it was only reduced to 55–89% of that in control cells. This was confirmed by electron microscopy. Furthermore, ricin endocytosed under conditions where uptake of transferrin and EGF was inhibited intoxicated cells to about the same extent as in control cells. This suggests that the acidification-resistant endocytosis is also important for the intoxication obtained with ricin under normal conditions.

To further investigate whether it is the acidification as such that is responsible for the inhibition of endocytosis from coated pits, we also acidified the cytosol by two other methods. In one case we incubated cells in a medium with pH 5.0 and added low concentrations of acetic acid, which can diffuse through the cell membrane and acidify the interior of the cell. Also in this case the endocytosis of transferrin and EGF was blocked, whereas there was little effect on the entry of ^{125}I-ricin.

The third method applied involves incubation of cells in a buffer containing the ionophores nigericin and valinomycin and isotonic concentrations of KCl. Owing to the ability of nigericin to exchange K^+ for H^+, the pH in the cytosol will become the same as that in the surrounding medium when the concentration of K^+ is the same at the two sides of the membrane. As shown in Figure 4, the endocytosis of transferrin and EGF was also blocked under these conditions, and the results confirm that strong inhibition of endocytosis is only obtained when the pH is below 6.5.

Earlier studies have indicated that cells have an internal pool of transferrin receptors and that various factors such as serum [24], insulin [25], and SH reagents [22] can induce a redistribution of these receptors resulting in an increased number at the cell surface. Furthermore, such a redistribution has been shown to be dependent on calcium in the medium. We noted that when cells were acidified by NH_4Cl deprivation or by addition of acetic acid more transferrin was bound than to control cells (Fig. 5). Scatchard analysis [26] showed that the increased binding was due to an increased number of receptors on the cell surface and not due to an increased affinity for transferrin. The increase was found to occur at pH values higher than those required to inhibit endocytosis, suggesting that the increase is due to an increased transport of receptors to the cell surface rather than to inhibition of internalization. Adelsberg and Al-Awqati [27] have found that low cytosolic pH can induce exocytosis of H^+ pumps, possibly by a related mechanism.

The data in Figure 5 also show that the increase in transferrin receptors at the cell surface could be abolished by removal of calcium from the medium, and that it was increased when the Ca^{2+} ionophore A23187 was added. It should be noted that addition of A23187 (10 μM) to Vero cells lowered the cytosolic pH value with about 0.07 pH units (not shown). However, this cannot be the reason for the observed effect since the absolute increase in receptor number was larger than that obtained at the highest NH_4Cl concentrations in the absence of the ionophore.

It is well established that the insulin-stimulated exocytosis of glucose transporters can be inhibited by trifluoperazine [28], and it has been suggested that calmodulin or another calcium-binding protein is involved in exocytosis. As shown in Figure 5,

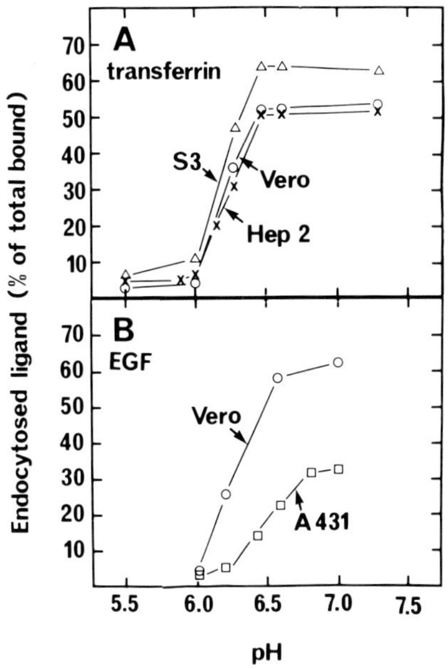

Fig. 4. Effect of pH-equilibration with nigericin and isotonic KCl on endocytic uptake of transferrin and EGF. Cells were growing in 24-well disposable trays to measure endocytic uptake of transferrin **(A)** and in Petri dishes (35 mM) for studies of EGF endocytosis **(B)**. After removal of the growth medium, 0.2 ml **(A)** or 0.5 ml **(B)** of 0.14 M KCl containing 2 mM $CaCl_2$, 1 mM $MgCl_2$, 5 μM nigericin, 10 μM valinomycin, and adjusted to the indicated pH values with Tris, was added. After a 5-min incubation at 37°C, ^{125}I-labeled transferrin (200 ng/ml, 38,600 cpm/ng or ^{125}I-EGF (7 ng/ml, 15,000 cpm/ng) was added, and the amount of endocytosed ligand was measured after 10 min (transferrin) or after 15 min (EGF).

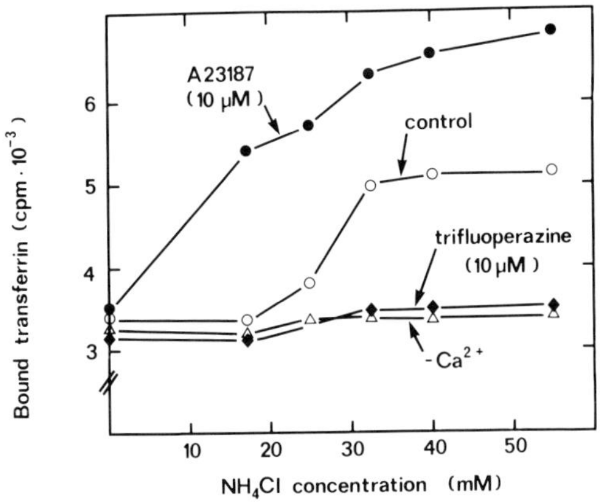

Fig. 5. Effect of calcium deprivation, trifluoperazine, and A23187 on the ability of low internal pH to induce increased binding of transferrin. Vero cells growing in 24-well disposable trays were incubated for 30 min at 37°C in Hepes medium (pH 7.0) with and without the indicated concentrations of NH_4Cl. The cells in some of the wells were incubated with trifluoperazine, while others were incubated in the absence of $CaCl_2$. The medium was then removed and 0.2 ml of buffer (0.14 M KCl, 1 mM $MgCl_2$, 2 mM $CaCl_2$, 1 mM amiloride, and 20 mM Hepes) was added. When indicated, $CaCl_2$ was omitted from the buffer. A23187 (10 μM) was added to some of the cells at this point. After an additional 5-min incubation the cells were chilled to 0°C, and ^{125}I-labeled transferrin (200 ng/ml, 38,600 cpm/ng) was added in a $CaCl_2$-containing medium. After a 1-hr incubation at 0°C, the cells were washed three times with cold medium, dissolved in 0.1 M KOH, and the amount of bound ^{125}I-transferrin was measured.

the increase in transferrin binding could also be abolished by trifluoperazine. Furthermore, W7 had a similar inhibitory effect, whereas the inactive analogue W5 had no effect. This supports the view that calmodulin or a similar regulatory protein is involved in the redistribution of transferrin receptors.

In the course of these studies, we noticed that low pH in the cytosol had a marked effect on the Golgi complexes that often became very vesiculated, with accumulation of numerous small, apparently coated, vesicular structures, which appeared to be connected to Golgi cisterns (Fig. 6). It is also possible that pinching off of coated vesicles in the Golgi apparatus is inhibited by the low cytosolic pH. However, further experiments are required to determine if this phenomenon is related to the effect on endocytosis.

In conclusion, the results presented here show that low pH in the cytosol inhibits endocytosis of transferrin and EGF in a reversible way, and that at low internal pH there is an increase in the number of transferrin receptors at the cell surface. Furthermore, as measured by studying endocytosis of ^{125}I-ricin, the results suggest that an alternative pathway of endocytosis exists, possibly involving uptake from noncoated areas of cell surface membrane and that the toxin ricin is able to utilize such a pathway on its way to the cytosol.

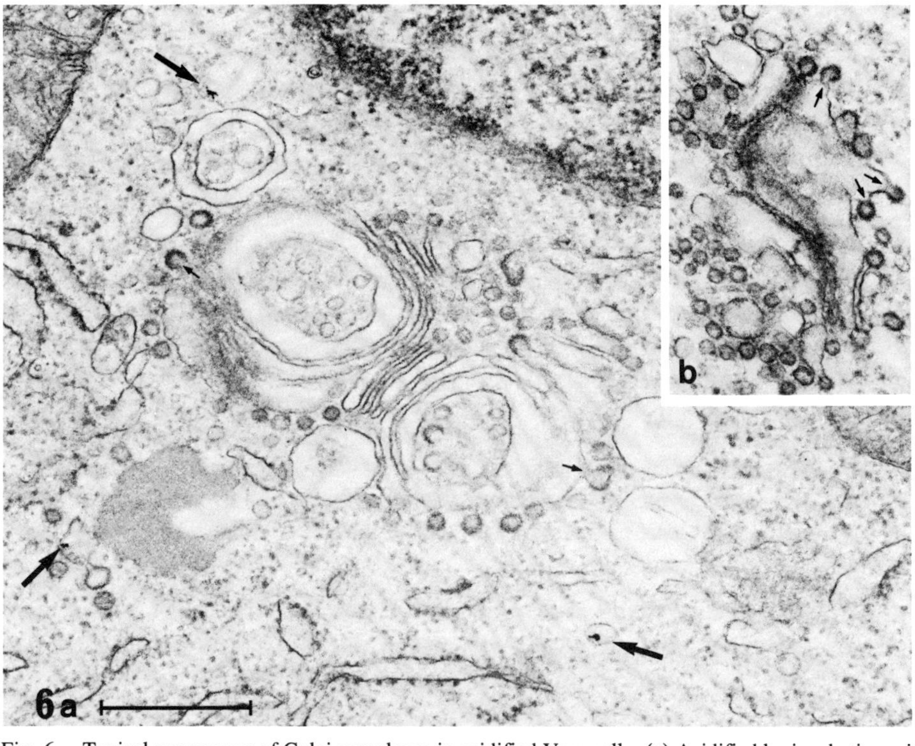

Fig. 6. Typical appearance of Golgi complexes in acidified Vero cells. **(a)** Acidified by incubation with nigericin and valinomycin in isotonic KCl pH 5.9. **(b)** Acidified by prepulsing with 25 mM NH_4Cl. **(a)** is from an experiment with Ri-Au (a few Ri-Au particles are seen in small vesicles; large arrows). **(b)** is from an experiment with Tf-Au. In both **(a)** and **(b)** vesiculated Golgi elements are distinct. Many small vesicles are connected to dilated cisterns (small arrows). a and b ×51,000.

ACKNOWLEDGMENTS

We wish to thank Tove Lie Berle, Jorunn Jacobsen, Keld Ottosen, and Kirsten Pedersen for excellent technical assistance. This work was supported by the NOVO Foundation.

REFERENCES

1. Moya M, Dautry-Varsat A, Goud B, Louvard D, Bouqet P: J Cell Biol 101:548, 1985.
2. Madshus IH, Sandvig K, Olsnes S, van Deurs B: J Cell Physiol 131:14, 1987.
3. Morris RE, Manhart MD, Saelinger CB: Infect Immun 40:806, 1983.
4. Morris RE, Gerstein AS, Bonventre PF, Saelinger CB: Infect Immun 50:721, 1985.
5. Montesano R, Roth J, Robert A, Orci L: Nature 296:651, 1982.
6. Madshus IH, Tønnessen TI, Olsnes S, Sandvig K: J Cell Physiol 131:6, 1987.
7. Dunn WA, Connolly TP, Hubbard AL: J Cell Biol 102:24, 1986.
8. Goldstein JL, Brown MS, Anderson RGW, Russell DW, Schneider WJ: Annu Rev Cell Biol 1:1, 1985.
9. Gordon P, Carpenter JL, Cohen S, Orci L: Proc Natl Acad Sci USA 75:5025, 1978.
10. Maxfield FR, Schlessinger J, Shechter Y, Pastan I, Willingham MC: Cell 14:805, 1978.
11. Olsnes S, Sandvig K: In Pastan I, Willingham MC (eds): "Endocytosis." New York: Plenum Press, 1985, pp 195–234.
12. Dautry-Varsat A, Ciechanover A, Lodish H: Proc Natl Acad Sci USA 80:2258, 1983.
13. Fraker PJ, Speck JC Jr: Biochem Biophys Res Commun 80:849, 1978.
14. Slot JW, Geuze HJ: J Cell Biol 90:533, 1981.
15. Horisberger M, Rosset J: J Histochem Cytochem 25:295, 1977.
16. Ciechanover A, Schwartz AL, Dauty-Varsat A, Lodish HF: J Biol Chem 258:9681, 1983.
17. Haigler HT, Maxfield FR, Willingham MC, Pastan I: J Biol Chem 255:1239, 1979.
18. Sandvig K, Olsnes S: Exp Cell Res 121:15, 1979.
19. Deutsch CJ, Holian A, Holian SK, Daniele RP, Wilson DF: J Cell Physiol 99:79, 1979.
20. van Deurs B, Ryde Pedersen L, Sundan A, Olsnes S, Sandvig K: Exp Cell Res 159:287, 1985.
21. van Deurs B, Tønnessen TI, Petersen OW, Sandvig K, Olsnes S: J Cell Biol 102:37, 1986.
22. Kaplan J, Ward DM, Wiley HS: J Cell Biol 101:121, 1985.
23. Moolenaar WH, Tertoolen LGJ, de Laat SW: J Biol Chem 259:7563, 1984.
24. Ward DM, Kaplan J: Biochem J 238:721, 1986.
25. Davis RJ, Corvera S, Czech MP: J Biol Chem 261:8708, 1986.
26. Scatchard G: Ann NY Acad Sci 51:660, 1949.
27. Adelsberg J, Al-Awqati Q: J Cell Biol 102:1638, 1986.
28. Shechter Y: Proc Natl Acad Sci USA 81:327, 1984.

Journal of Cellular Biochemistry 36:261–273 (1988)
Molecular Biology of Intracellular Protein
Sorting and Organelle Assembly 57–69

Posttranslational Modification of *ras* Proteins: Detection of a Modification Prior to Fatty Acid Acylation and Cloning of a Gene Responsible for the Modification

F. Tamanoi, E.C. Hsueh, L.E. Goodman, A.R. Cobitz, R.J. Detrick, W.R. Brown, and A. Fujiyama

Department of Biochemistry and Molecular Biology, The University of Chicago, Chicago, Illinois 60637

Products of *ras* genes are synthesized as precursors in the cytosol and transported to the plasma membrane by a process which involves posttranslational modification by fatty acid. In this paper, we present evidence for the occurrence in the cytosol of an intermediate modification of *ras* proteins prior to the fatty acid acylation. The modification is detected by a slight shift in the mobility of the protein on SDS polyacrylamide gel. The fatty acid acylation does not contribute to this mobility shift. This modification is affected by the *dpr1* mutation which has recently been shown to affect the processing of yeast *RAS* proteins. To further characterize the nature of the modification event, we have cloned *DPR1* gene from the DNA of *Saccharomyces cerevisiae*. The gene is actively transcribed in yeast cells producing mRNA of approximately 1.6 kb. Genes related to the *DRP1* appear to be present in a distantly related yeast, *Schizosaccharomyces pombe* as well as in guinea pig and human cells.

Key words: ***Schizosaccharomyces pombe, Saccharomyces cerevisiae*, subcellular localization, *DPR1* gene, processing of *ras* protein**

Products of *ras* genes are synthesized on free polysomes and are localized at the inner surface of the plasma membrane [1,2]. How *ras* proteins travel to the plasma membrane poses an interesting problem in protein transport, since they appear to be different from typical secretory proteins in that no apparent signallike sequences can be found in these proteins. Instead they are fatty acid acylated and it is possible that this fatty acid serves as an anchor to the plasma membrane by virtue of its hydrophobic properties [3–5].

The association of *ras* proteins to the plasma membrane is essential for their function, since *ras* proteins which are not associated with the plasma membrane

Received February 5, 1987; revised and accepted July 28, 1987.

cannot transform NIH/3T3 cells [4,5]. Studies employing in vitro mutagenesis of *ras* genes have shown that a cysteine located at the fourth position from the carboxy terminus is involved in the fatty acid acylation of *ras* proteins. This cysteine is followed by two hydrophobic amino acid residues, and the sequence CysAAX (where A is an aliphatic amino acid and X is the C-terminal amino acid) is conserved in all *ras* proteins [6–9].

Characterization of the processing of the products of *ras* gene homologues in *Saccharomyces cerevisiae* has provided an insight into these post-translational modification processes. We have shown [10] that the protein products of the homologous genes, *RAS1* and *RAS2*, are synthesized as precursor proteins. In the first modification step, these precursor forms are converted to intermediate forms which migrate slightly faster on SDS polyacrylamide gel electrophoresis than their precursor forms. Fatty acid acylation of these intermediate forms is the second step of modification for *RAS* proteins in yeast and the majority of fatty-acid-acylated *RAS* proteins are localized in the plasma membrane. We have proposed the following scheme for the posttranslational modification: precursor *RAS* → intermediate *RAS* → fatty-acid-acylated *RAS* [10,11]. The implication of this scheme for the processing of *ras* proteins in general is that there should be an identifiable modification of *ras* proteins that necessarily precedes fatty acid acylation. Occurrence of a modification distinct from fatty acid acylation has also been suggested from the characterization of H-*ras* proteins produced in NIH/3T3 cells [12].

In the course of our attempt to express the mammalian H-*ras* gene in yeast, we have noticed that a processing very similar to the one observed with yeast *RAS* proteins takes place with the H-*ras* proteins and that the processing of H-*ras* proteins is slower than that of yeast *RAS* proteins. These observations, the latter in particular, suggested to us that a form of *ras* proteins representing an intermediate in the posttranslational modification can be clearly identified by using this system. In this paper, we describe processing of H-*ras* proteins produced in yeast and show that the processing of *ras* proteins, indeed, occurs in two steps.

The recent isolation of yeast mutants *dpr1* [11] and *ram* [13], which are defective in the processing of *ras* proteins, allows us the opportunity to study this system of modification for *ras* proteins by powerful genetic methods. In the *dpr1* mutant, yeast *RAS2* proteins are accumulated as precursors in the cytosol, suggesting that the *dpr1* mutation primarily affects the conversion of a precursor form to an intermediate form [11]. One interesting phenotype of the *dpr1* and *ram* mutants is a sterility specific of *MATa* cells [11,13]; *MATa* cells having the mutation cannot mate with *MATα* cells. This is apparently due to a defect in the production of mature *a*-factor which is required for the mating with *MATα* cells [13] (R. Sterne and J. Thorner, personal communication). In fact, intracellular accumulation of the unprocessed *a*-factor can be detected in the *dpr1* mutant cells (R. Sterne and J. Thorner, personal communication). Therefore, the intermediate modification that occurs on *ras* proteins prior to their fatty acid acylation also plays a major role in the maturation of the *a*-factor. To gain further insights into this modification event, we have cloned the *DPR1* gene responsible for the modification. It appears to be a single gene in yeast and is actively transcribed. Survey for the presence of genes related to the *DPR1* in diverse organisms suggests that the gene is conserved during evolution.

MATERIALS AND METHODS

Yeast Strains and Materials

The following *S. cerevisiae* strains were used. UC100 (*MATa leu2 trp1 ura3 pep4 prb*) and G441-2D(*MATα leu1*) were obtained from R. Kostriken (Cold Spring Harbor Laboratory) and T. Toda (Cold Spring Harbor Laboratory), respectively. The strains KMY5-2A(*MATα dpr1 his3 leu2 ura3*) and HR13(*MATa dpr1 leu2 his3 ura3 trp1 ade8 can1 RAS2*^{val19}) were obtained from K. Matsumoto (DNAX Research Institute of Molecular and Cellular Biology). [^{35}S]methionine (1,130 Ci/mmol), [^{3}H]palmitic acid (55 Ci/mmol), and [$^{32}\alpha$-P]dCTP (3,000 Ci/mmol) were purchased from Amersham.

Construction of Plasmid DNA

Construction of pYG-H-*ras*, a vector to express H-*ras* gene in yeast, was accomplished by placing the H-*ras* gene under the control of glyceraldehyde 3-phosphate dehydrogenase promoter. For this purpose, the H-*ras* gene was cut out from pOFSP-2 [14] (provided by M. Wigler, Cold Spring Harbor Laboratory) by the digestion with SalI and BanII. After filling the ends with Klenow fragment, the fragment was blunt end ligated to the SalI site of pYG10 (provided by K. Matsubara, Osaka University) which had also been blunt ended by filling the ends. The orientation of the insert was examined by restriction digestions. YEp24-DPR1 was constructed by inserting a BamHI to SalI fragment of *DPR1* into YEp24 DNA. The YEp51-RAS2 plasmid has been described previously [15].

Metabolic Labeling and Immunoprecipitation

Labeling of yeast cells with [^{35}S]methionine or [^{3}H]palmitic acid, fractionation into soluble and membrane fractions, immunoprecipitation of *ras* proteins with monoclonal antibody 259, and SDS polyacrylamide gel electrophoresis were carried out as previously described [10].

Cloning of *DPR1* Gene

For the cloning of *DPR1* gene, a library of *S. cerevisiae* genomic DNA was used. This library (provided by R. Elder, University of Chicago) was constructed by inserting Sau3A partial digests into a BamHI site of YCp19. The DNA was transformed into a temperature-sensitive *dpr1* mutant, HR13, according to Beggs [16] and screened for the ability of transformants to grow at a high temperature (35°C). Plasmid DNAs from the transformants were isolated and recovered by transforming into *E. coli* according to Hanahan [17]. DNAs isolated were used to transform the HR13 cells in order to confirm their complementing ability.

DNA and RNA Analyses

Total yeast RNA was isolated from log-phase cells by breaking cells in the presence of guanidine isothiocyanate. Briefly, cells were suspended in 4 M guanidine isothiocyanate/2 M 2-mercaptoethanol/0.1 M Tris HCl (pH7.5)/5 mM EDTA and broken with glass beads. After removing unbroken cells and glass beads, RNA was purified by CsCl density-gradient centrifugation. Poly(A)$^{+}$ RNA was isolated by using an oligo-dT cellulose column. The RNA samples were size fractionated on an 0.8% agarose/2.2 M formaldehyde gel, transferred to nitrocellulose filter, and hy-

intermediate form, suggesting that the fatty acid acylation does not alter the mobility of *ras* proteins on SDS polyacrylamide gel electrophoresis.

The relationship between these three forms of *ras* protein can be deduced from pulse-chase experiments shown in Figure 2C. After a 10-min pulse with [^{35}S]methionine, most of the radioactivity was detected in the precursor form. This is in marked contrast to the processing of yeast *RAS* proteins, in which virtually no precursor forms are detected after a 2-min pulse [10]. When a chase with unlabeled methionine was performed, radioactivity in the intermediate form increased, and eventually there was radioactivity in the fatty-acid-acylated form. These results suggest that the precursor form is converted to the intermediate form, which then becomes fatty acid acylated.

dpr1 Mutation Affects the Intermediate Modification Which Precedes the Fatty Acid Acylation

Recently, a mutant of yeast, *dpr1*, defective in the processing of yeast *RAS* proteins has been isolated. Figure 3 shows the effect of the mutation on the processing of H-*ras* proteins. The plasmid pYG-H-ras was introduced into the *dpr1* mutant cells and the transformants were labeled with [^{35}S]methionine. The *ras* proteins were immunoprecipitated with the monoclonal antibody 259. As can be seen in Figure 3A lane 2, most of the radioactivity in the *dpr1* mutant was detected in the 22,000-dalton

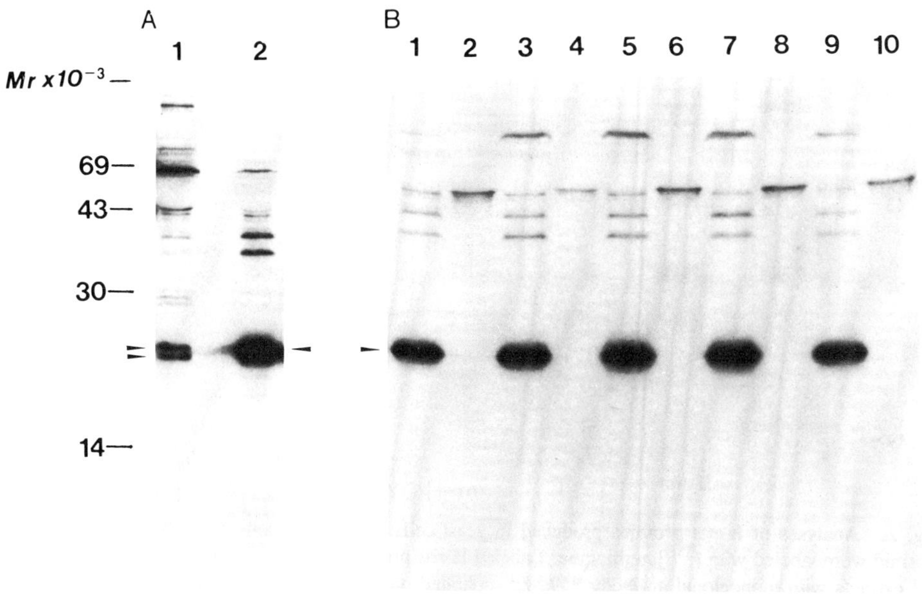

Fig. 3. Effect of the *dpr1* mutation on the processing of H-*ras* proteins produced in yeast cells. **A:** UC100 cells (lane 1) and KMY5-2A cells (lanes 2) were transformed with pYG-H-ras plasmid and labeled with [^{35}S]methionine. Labeled H-*ras* proteins were immunoprecipitated and separated as described in Figure 2. **B:** Pulse-chase analysis of H-*ras* proteins in *dpr1* cells. KMY5-2A cells carrying pYG-H-ras were pulse labeled with [^{35}S]methionine and chased as described in Figure 2C. Pulse labeling for 10 min (lanes 1 and 2), chase for 30 min (lanes 3 and 4), 60 min (lanes 5 and 6), 120 min (lanes 7 and 8), and 240 min (lanes 9 and 10). Labeled cells were separated into soluble (lanes 1, 3, 5, 7, and 9) and membrane fractions (lanes 2, 4, 6, 8, and 10). Immunoprecipitation and SDS/PAGE were performed as described in Figure 2.

precursor *ras* protein. In contrast to this, both the precursor and processed *ras* proteins were detected in parental cells (Fig. 3A, lane 1). The precursor *ras* proteins in the *dpr1* mutant were not processed even after a very lengthy chase with cold methionine (Fig. 2B). Fractionation of extracts of the *dpr1* mutant into soluble and membrane fractions revealed that the H-*ras* proteins were present almost exclusively in the 22,000-dalton form in the soluble fraction. These results suggest that the *dpr1* mutation affects the first modification step; conversion of a precursor form to an intermediate form. In addition, the mutation affects localization of the protein.

Cloning of *DPR1* Gene

Further investigation into the nature of the intermediate modification of *ras* proteins would be greatly facilitated by the identification of a gene responsible for the modification. We cloned the *DPR1* gene by exploiting temperature-sensitive growth of the *dpr1* mutant. Transformation of the *dpr1* mutant HR13 with DNAs from genomic library of *S. cerevisiae* resulted in the isolation of five plasmids (YCp-DPR 12, 21, 30, 41, 311) which complemented the *ts* phenotype. Yeast DNA fragments contained in these plasmids are shown in Figure 4. A common region of 4.5 kb is found in these plasmids.

Another phenotype of the *dpr1* mutant is an *a* sterile phenotype. To demonstrate that the cloned gene also complements this *a* sterile phenotype, each of the five YCp-DPR plasmids described above was introduced into HR13 which is *MATa dpr1*. The transformants of HR13 carrying YCp-DPR were able to mate with G441-2d(*MATα leu1*) cells. On the other hand, HR13 cells not carrying the YCp-DPR plasmid were not able to mate with G441-2D cells. Thus, the plasmid YCp-DPR confers *MATa dpr1* cells the ability to mate with *MATα* cells.

We next examined whether the *DPR1* clones are capable of complementing the deficiency of *ras* protein processing in the *dpr1* cells. A *dpr1* mutant KMY5-2A was transformed with a plasmid YEp51-RAS2 to overproduce yeast *RAS2* protein. The transformants were then transformed with YCp-DPR and selected at 35°C. The two

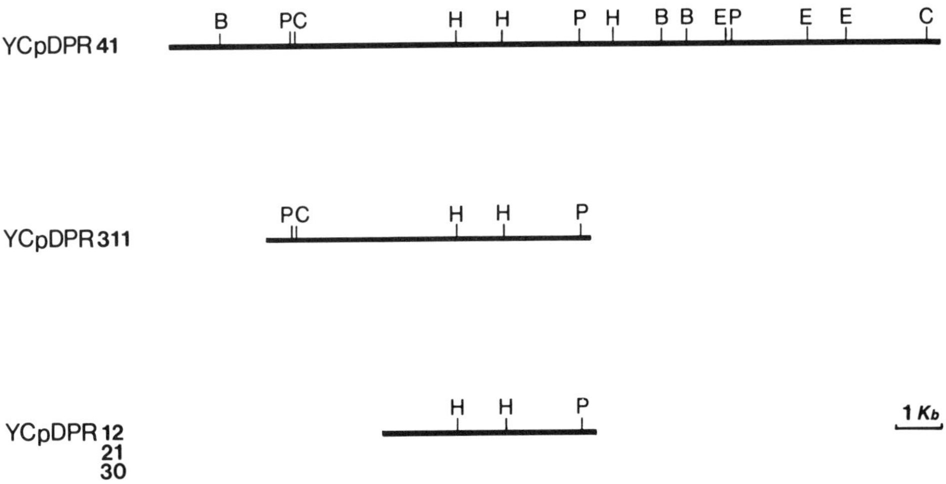

Fig. 4. Structure of cloned *DPR1* fragments. Cleavage sites by restriction endonucleases on cloned DNA fragments are shown. B, BamHI, C, ClaI, E, EcoRI, H, HpaI, P, PstI.

plasmids were maintained on a medium lacking both leucine and uracil. Processing of *RAS2* protein was examined by immunoprecipitating [^{35}S]methionine-labeled *RAS2* protein with the monoclonal antibody 259. As can be seen in Figure 5, processing of precursor *RAS2* proteins was observed in *dpr1* cells transformed with the *DPR1* clones (lanes 2–4). However, in the *dpr1* cells transformed with the vector but lacking the *DPR1* insert, only precursor proteins were detected (Fig. 5, lane1). Thus, the processing of RAS2 protein is complemented by the *DPR1* clones. In an entirely

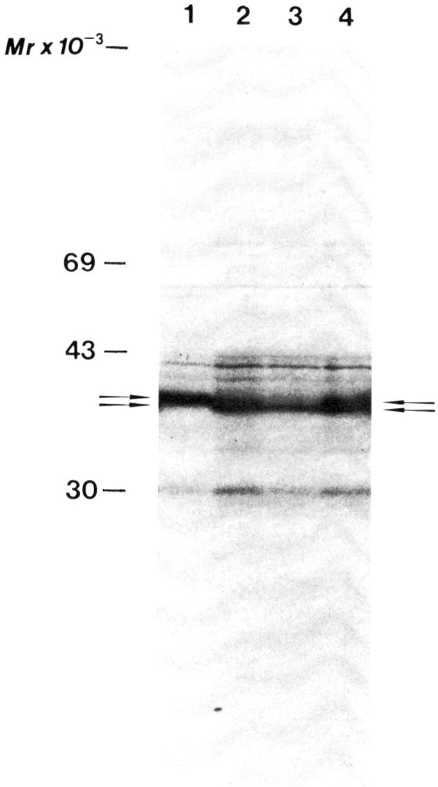

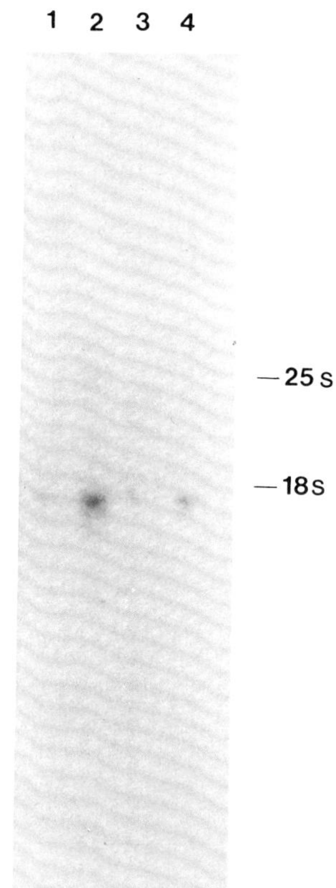

Fig. 5. Complementation of the *dpr1* phenotype by the cloned *DPR1* plasmids. HR13 cells carrying the YEp51-RAS2 plasmid were transformed with YCp19 (lane 1), YCp-DPR12 (lane 2), YCp-DPR41 (lane 3), and YCp-DPR311 (lane 4) DNA. Doubly transformed cells were labeled with [^{35}S]methionine and the processing of yeast *RAS2* proteins was examined as described in Figure 2.

Fig. 6. Blot-hybridization analysis of *DPR1* expression in yeast cells. Total and poly(A)$^+$ RNA were size fractionated on 0.8% agarose/2.2 M formaldehyde gel, transferred to nitrocellulose filter, and hybridized with the *DPR1* probe at 42°C overnight in 5 × SSC/5×Denhardt solution/20 mM sodium phosphate (pH 6.9)/1% SDS/50% formamide. The filter was washed twice for 5 min at room temperature in 2×SSC, followed by a 1-hr wash at 65°C in 0.1×SSC/0.1 M sodium phosphate (pH6.9)/0.05% SDS/1×Denhardt solution, then followed by another 1-hr wash at 65°C in 1×SSC/0.01 M sodium phosphate (pH 6.9)/0.05% × SDS/1×Denhardt solution, and finally washed with 0.1×SSC. Lane 1: 11 μg of poly(A)$^+$ RNA from UC100 cells carrying a plasmid YEp24. Lane 2: 11 μg of poly(A)$^+$ RNA from UC100 cells carrying YEP24-DPR1. Lane 3: 30 μg of total RNA from UC100 cells carrying YEp24. Lane 4: 30 μg of total RNA from UC100 cells carrying YEp24-DPR1. Markers are yeast ribosomal RNAs.

analogous procedure, we have also shown that the *DPR1* clones complement the processing defect of H-*ras* proteins in yeast (data not shown).

Detection of *DPR1* Transcript

Transcription of the *DPR1* gene was assessed by Northern hybridization of yeast RNA with the cloned *DPR1* DNA. As shown in Figure 6, a single band with an approximate size of 1.6 kb was detected. The level of the *DPR1* mRNA is elevated in cells carrying YEp24-DPR1, a multicopy plasmid containing the *DPR1* gene (compare lanes 3 and 4). Similar results were obtained with poly(A)$^+$ RNA, indicating that the *DPR1* mRNAs are polyadenylated. Therefore, the *DPR1* gene is actively transcribed in yeast cells.

DPR1 Gene Appears to Be Conserved During Evolution

The availability of the cloned gene enabled us to search for the presence of genes related to the *DPR1* in various organisms. The cloned gene was first hybridized against genomic DNA from *S. cerevisiae*. As can be seen in Figure 7A, a single band was always found regardless of which endonuclease was used for digestions, suggesting that the *DPR1* gene is a single gene in *S. cerevisiae*, in agreement with other genetic analyses. We next investigated whether genes related to the *DPR1* are present in other organisms. Figure 7B shows the results of hybridization analyses using genomic DNA from *S. pombe*. Clear bands were detected when the hybridization and washing were carried out under nonstringent as well as stringent hybridization conditions. These bands did not correspond to abundant DNA detected by ethidium bromide staining. Figure 8 shows further analyses using guinea pig and human DNAs. In these cases also, we detected bands which did not correspond to repetitive DNA bands seen with ethidium bromide staining. These bands were detected even when the washing was carried out under stringent conditions. These results raise the possibility that genes related to the *DPR1* are present in a distantly related yeast, *S. pombe*, as well as in higher eukaryotes (guinea pig and human cells).

DISCUSSION

The modification of *ras* proteins leading to their acylation by fatty acids involves two distinct modification events: conversion of precursor forms to intermediate forms and fatty acid acylation of the intermediate forms. Our argument that there is an intermediate modification of *ras* proteins prior to their fatty acid acylation rests on the following two pieces of evidence. First, a form distinct from either the precursor or the fatty-acid-acylated form can be detected in the soluble fraction. This molecule displays characteristics expected for an intermediate of processing by pulse-chase experiment. Second, since the intermediate molecule migrates at approximately the same position as the fatty-acid-acylated molecule, it appears that the effect of fatty acid acylation on its mobility on SDS polyacrylamide gel is little, if any. Presence of a modification distinct from fatty acid acylation has also been suggested from the analyses of H-*ras* protein produced in NIH/3T3 cells [12]. However, in two reports that have appeared concerning processing of H-*ras* proteins in yeast [13,20], only precursor and fatty-acid-acylated forms were detected. The reason for the discrepancy is unclear at the moment.

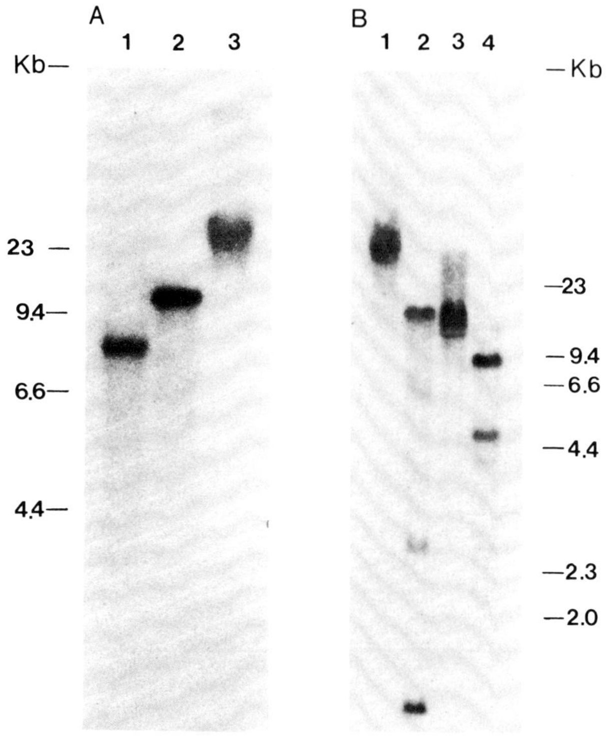

Fig. 7. Southern hybridization analyses of *S. cerevisiae* DNA and *S. pombe* DNA. **A:** DNA from *S. cerevisiae* (3 μg/lane) was digested with either BamHI (lane 1) or EcoRI (lane 2) and hybridized with labeled *DPR1* fragment under stringent conditions, hybridized at 42°C overnight in 50% formamide/ 5×SSC/5×Denhardt solution/20 mM sodium phosphate (pH 6.5)/0.1% SDS, washed twice at room temperature for 5 min each with 2 × SSC followed by three more washes at 65°C for 1 hr each with 2 × SSC/0.1 × Denhardt solution/0.1 M sodium phosphate, (pH 7.4)/0.1% SDS. Lane 3: No digestion. A Hind III digest of lambda phage DNA was used as a size marker. **B:** DNA from *S. pombe* (20 μg/ lane) was digested with EcoRI (lane 2), BamHI (lane 3), and HindIII (lane 4). Hybridization was performed under nonstringent condition; hybridized at 37°C overnight in 5 × SSC/5 × Denhardt solution/20 mM sodium phosphate (pH 6.5)/0.1% SDS, washed twice at room temperature for 5 min each with 2 × SSC followed by three more washes at 56°C for 1 hr each with 2 × SSC/0.1 × Denhardt solution/0.1 M sodium phosphate (pH 7.4)/0.1% SDS. Lane 1: No digestion.

The two modification events seen with *ras* proteins are presumably related. This is suggested from the following two independent observations. First, the *dpr1* mutation blocks the first modification (conversion of a soluble precursor form to a soluble intermediate form), and the amount of fatty-acid-acylated *ras* proteins is greatly reduced in the mutant [11]. This result implies that the first modification is an obligatory step for the second modification (fatty acid acylation). Second, the alteration of cysteine to serine at the CysAAX sequence, a presumed fatty acylation site, also affects the first modification [4], suggesting that the two modifications occur at the same or nearby sites. Our analyses of the amino terminal sequences of the precursor and mature yeast *RAS2* proteins suggest that no modifications occur at the amino terminus. Both proteins were lacking the first methionine predicted from the DNA sequence, but the second amino acid (proline) was present in both cases. Furthermore, formic acid treatment of the two proteins (which cleaves the *RAS2*

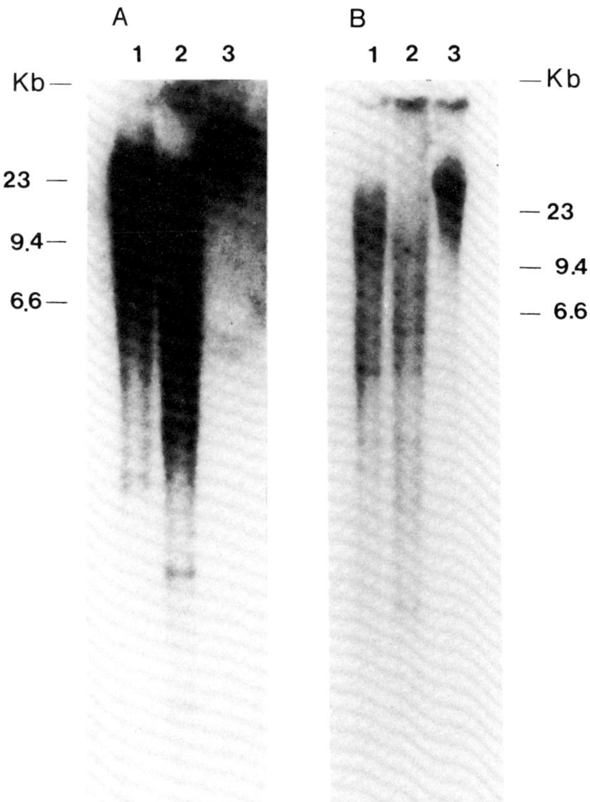

Fig. 8. Southern hybridization analyses of guinea pig and human DNAs; 20 μg of guinea pig DNA (**A**) or human DNA (**B**) was digested with BamHI (lane 1) or EcoRI (lane 2) and run on a 0.8% agarose gel. After transferring DNA onto nitrocellulose filter, the DNA was hybridized with labeled *DPR1* probe. Hybridization was carried out in 5 × SSC/5 × Denhardt solution/20 mM sodium phosphate (pH 6.5)/ 0.1% SDS at 37°C overnight, washed twice at room temperature for 5 min each with 2 × SSC/0.1 × Denhardt solution/0.1 M sodium phosphate (pH 7.4)/0.1% SDS followed by two more washes at 50°C for 1 hr each in the same buffer, and further washed twice more at 60°C for 1 hr each in the same buffer. Lane 3 shows undigested DNA. A HindIII digest of lambda phage DNA was used as a size marker.

protein between asp39 and pro40 from the amino terminus) produced carboxy terminal fragments which still retained the mobility shift (unpublished).

The exact chemical nature of the modification which precedes fatty acid acylation has not been elucidated. However, recent reports on the structure of mating factors of yeast shed some light on this problem. The cDNA sequence of the *a*-factor of *S. cerevisiae* suggests that the *a*-factor is produced as a precursor of 36 or 38 amino acids which contains a sequence CysValIleAla at the carboxy terminus [21]. The mature *a*-factor, however, contains 12 amino acid residues, and a cysteine which is alkylated is present at the carboxy terminus [22]. The mating pheromone of basidiomycetous yeast, *Rhodosporidium toruloides*, is an undecapeptide having S-farnesylated cysteine at the carboxy terminus [23]. The sequence of the cloned gene revealed the presence of three amino acids following the cysteine [24]. These observations are striking in that the terminal three amino acids are removed in both cases. This result raises the possibility that the carboxy terminal three amino acids in the *ras* proteins are removed as well before the fatty acid acylation takes place. Direct determination of the carboxy terminal sequence of the *ras* proteins is needed to establish this point.

To facilitate the characterization of the first modification event, we have cloned the *DPR1* gene. The gene cloned by its ability to complement the temperature-sensitive phenotype of the *dpr1* mutant is capable of reversing the *a*-sterile phenotype as well as restoring the ability to process precursor *ras* proteins in the *dpr1* mutant. Thus, all these phenotypes are due to a mutation in a single gene. Powers et al. have cloned a gene, *RAM*, which also affects the processing of *ras* proteins and the *a*-factor precursor [13]. Comparison of the restriction maps of the *RAM* gene and the *DPR1* gene indicates that the two genes are identical. From the Southern analyses, it appears that there is only one *DPR1* gene, although presence of a gene with a limited homology has not been ruled out. The *DPR1* gene is actively transcribed in yeast cells giving rise to a mRNA of approximately 1.6 kb.

Availability of the cloned gene enabled us to investigate whether genes related to the *DPR1* are present in other organisms. Southern hybridization experiments with genomic DNAs suggest that a related gene is present in the distantly related yeast *S. pombe*. By similar hybridization experiments using genomic DNAs from guinea pig and human, we have obtained indications that genes related to the yeast *DPR1* might be present in these higher eukaryotes as well. The *ras* genes are present in diverse species ranging from yeast to human. The genes involved in the modification of the products of the *ras* genes are also conserved during evolution.

ACKNOWLEDGMENTS

We thank Drs. E. Goldwasser, T. Petes, and E. Westbrook for critical reading of the manuscript. This work was supported by NIH grant RO1 CA 41996 and in part by the Louis Block Fund, The University of Chicago.

REFERENCES

1. Shih TY, Weeks MO, Gruss P, Dhar R, Oroszlan S, Scolnick EM: J Virol 42:253, 1982.
2. Willingham MC, Pastan I, Shih TY, Scolnick EM: Cell 19:1005, 1980.
3. Sefton BM, Trowbridge IS, Cooper JA, Scolnick EM: Cell 31:465, 1982.
4. Willumsen BM, Norris K, Papageorge AG, Hubbert NL, Lowy DR: EMBO J 3:2581, 1984.
5. Weeks MO, Hager GL, Lowe R, Scolnick EM: J Virol 54:586, 1985.
6. Taparowsky E, Shimizu K, Goldfarb M, Wigler M: Cell 34:581, 1983.
7. Schejter ED, Shilo BZ: EMBO J 4:407, 1985.
8. Reymond CD, Gomer RH, Mehdy MC, Firtel RA: Cell 39:141, 1984.
9. Powers S, Kataoka T, Fasano O, Goldfarb M, Strathern J, Broach J, Wigler M: Cell 36:607, 1984.
10. Fujiyama A, Tamanoi F: Proc Natl Acad Sci USA 83:1266, 1986.
11. Fujiyama A, Matsumoto K, Tamanoi F: EMBO J 6:223, 1987.
12. Buss JE, Sefton BM: Mol Cell Biol 6:116, 1986.
13. Powers S, Michaelis S, Broek D, Santa Anna-A S, Field J, Herskowitz I, Wigler M: Cell 47:413, 1986.
14. Kataoka T, Powers S, Cameron S, Fasano O, Goldfarb M, Broach J, Wigler M: Cell 40:19, 1985.
15. Tamanoi F, Walsh M, Kataoka T, Wigler M: Proc Natl Acad Sci USA 81:6924, 1984.
16. Beggs JD: Nature 275:104, 1978.
17. Hanahan D: J Mol Biol 166:557, 1983.
18. Maniatis T, Fritsch EF, Sambrook J: "Molecular Cloning: A Laboratory Manual." Cold Spring Harbor, New York: Cold Spring Harbor Laboratory, 1982.

19. Cryer DR, Eccleshall R, Marmur J: "Methods in Cell Biology," 12. New York: Academic Press, 1975, pp 39–44.
20. Clark SG, McGrath JP, Levinson AD: Mol Cell Biol 5:2746, 1985.
21. Brake AJ, Brenner C, Najarian R, Laybourn P, Merryweather J: "Protein Transport and Secretion." Cold Spring Harbor, New York: Cold Spring Harbor Laboratory, 1985, pp 103–108.
22. Betz R, Crabb JW, Meyer HE, Wittig R, Duntze W: J Biol Chem 262:546, 1987.
23. Kamiya Y, Sakurai A, Tamura S, Takahashi N, Abe K, Tsuchiya E, Fukui S, Kitada C, Fujino M: Biochem Biophys Res Commun 83:1077, 1978.
24. Akada R, Minomi K, Yamashita I, Miyakawa T, Fukui S: Agric Biol Chem (Jpn) 51:1211, 1987.
25. Holland JP, Holland MJ: J Biol Chem 254:9839, 1979.

Journal of Cellular Biochemistry 36:275–287 (1988)
Molecular Biology of Intracellular Protein Sorting and Organelle Assembly 71–83

Functional Analysis of Mitochondrial Protein Import in Yeast

Scott M. Glaser, Cynthia E. Trueblood, Lori K. Dircks, Robert O. Poyton, and Michael G. Cumsky

Department of Molecular Biology and Biochemistry, University of California, Irvine, Irvine, California (S.M.G., M.G.C.), Department of Molecular, Cellular, and Developmental Biology, University of Colorado, Boulder, Colorado 80309 (C.E.T, L.K.D., R.O.P.)

In order to facilitate studies on protein localization to and sorting within yeast mitochondria, we have designed an experimental system that utilizes a new vector and a functional assay. The vector, which we call an LPS plasmid (for leader peptide substitution), employs a yeast *COX5a* gene (the structural gene for subunit Va of the inner membrane protein complex cytochrome *c* oxidase) as a convenient reporter for correct mitochondrial localization. Using in vitro mutagenesis, we have modified *COX5a* so that the DNA sequences encoding the wild-type subunit Va leader peptide can be precisely deleted and replaced with a given test sequence. The substituted leader peptide can then be analyzed for its ability to direct subunit Va to the inner mitochondrial membrane (to target and sort) by complementation or other in vivo assays. In this study we have tested the ability of several heterologous sequences to function in this system. The results of these experiments indicate that a functional leader peptide is required to target subunit Va to mitochondria. In addition, leader peptides, or portions thereof, derived from proteins located in other mitochondrial compartments can also be used to properly localize this polypeptide. The results presented here also indicate that the information necessary to sort subunit Va to the inner mitochondrial membrane does not reside in the leader peptide but rather in the mature subunit Va sequence.

Key words: cytochrome oxidase, subunit Va, in vitro mutagenesis

Recent results from several laboratories have clearly established the critical role of the amino terminal leader peptide (or presequence) in the selective targeting of proteins to, and into the mitochondrion [for recent reviews see references 1 and 2]. Numerous studies, most of them involving chimeric gene fusions, have demonstrated that the leader peptide is necessary and often sufficient for mitochondrial targeting [3,4]. In some cases, it has been shown that the leader peptide contains sequences

Abbreviations used: kb, kilobase pair; dNTPs, deoxynucleotide triphosphates; SDS, sodium dodecyl sulfate; TRP^+, tryptophan positive; TMPD, tetramethyl-p-phenylenediamine.

Received June 11, 1987; revised and accepted August 17, 1987.

necessary for correct intramitochondrial localization (sorting) as well [5]. Recently, it has been proposed that many leader peptides on imported proteins have the ability to form amphipathic helices which might be a common structural feature important in their function [6,7].

While much has been learned about mitochondrial leader peptides and their role in protein import, several important questions remain answered. Are specific amino acid residues required for targeting or is secondary structure alone sufficient? Does all of the information required to target and sort an imported protein reside in its leader peptide or does the mature polypeptide contribute? How many ancillary factors (e.g., receptors, soluble factors) [8,9] are required for import, and do they interact with the leader peptide?

To address these and other questions, we have developed a new experimental system. Central to it is the yeast gene *COX5a*, which encodes subunit Va of the mitochondrial inner membrane protein complex cytochrome *c* oxidase. The *COX5a* gene, one of two nuclear genes specifying divergent forms of subunit V in yeast [10,11], is typical of the majority of yeast nuclear genes encoding mitochondrial proteins in that it is initially synthesized in precursor form. The preprotein contains a basic 20-residue leader peptide which is proteolytically removed during or after import into the organelle [12]. Gene disruption studies which have inactivated one or both *COX5* genes have established that a form of subunit V is required for cytochrome oxidase activity and hence for cellular respiration [11,13]. Thus, in yeast strains lacking these genes, there is no detectable respiration [11,13]. Such strains are unable to grow on nonfermentable substrates like glycerol or lactate, although wild-type respiration can be restored by the introduction of a single copy of *COX5a* or by multiple copies of *COX5b* [11,13].

In this study, we have constructed a plasmid in which the wild-type *COX5a* gene has been modified such that the sequences encoding the Va leader peptide can be removed and replaced with a heterologous sequence of choice. When a recombinant plasmid is transformed into a yeast host deleted for both subunit V genes, the ability of the new sequence to direct the attached protein to the inner mitochondrial membrane can be assayed by screening for the ability of transformants to respire. We have thus far tested the ability of several different sequences to target subunit Va in this system. We report here the results of these initial studies.

MATERIALS AND METHODS

Strains and Growth Media

The wild-type *Saccaromyces cerevisiae* strains used in this study were D273-10B (*mat*α, ATCC 24657) and JM43 (*mat*α *leu2-3 leu2-112 his4-580 ura3-52 trp1-289*), and both have been described previously [11]. The strain used for complementation studies was JM43-GD5ab (*mat*α *leu2-3 leu2-112 his4-580 ura3-52 trp1-289 cox5a*Δ*::URA3 cox5b::LEU2*). This strain is a derivative of JM43 in which the chromosomal copies of the *COX5a* and *COX5b* genes were sequentially disrupted with the *URA3* and *LEU2* genes, respectively [13]. When not harboring an extrachromosomal plasmid it has no detectable respiration and thus will not grow on any nonfermentable substrates [13]. Yeast strains were routinely grown at 30°C in selective or nonselective media as described [11]. Respiratory proficiency was tested on YPEG or YP lactate media [14].

E. coli strains used for plasmid propagation were HB101, RR1, or DH1 [11]. For phage cloning and propagation, JM101 or a *recA*$^-$ derivative of JM103 was used [11]. *E. coli* strains were grown at 37°C in LB or YT medium [11]. When necessary, transformants were selected in the above medium containing 100 μg/ml ampicillin.

Plasmid Construction

The plasmid YCpLPS-5a (Fig. 1) was constructed in several steps, as follows. A 3.4-kb EcoR1 fragment containing the *COX5a* structural gene and over a kilobase of both 5′ and 3′ flanking sequence, was isolated from the plasmid YEp13-552 [11] and cloned in the M13 phage vector mp19 [15]. Single-stranded template DNA prepared from JM103 cells carrying this phage (named mp19-5aEco) was then used as a template for in vitro mutagenesis. A synthetic oligonucleotide (5′-GCTCAA-ACGCATGCTCTT-3′) was used to create a unique SphI restriction site near the amino-terminal end of the mature subunit Va sequence (Fig. 2). This mutation changes codon 3 of the mature sequence from ACA to ACG but does not alter the amino acid specified at that position (thr). The mutagenesis procedure followed was essentially the two-primer method of Zoller and Smith [16] except that the ratio of primers to template was 1:1, the annealing was for 5 min at 65°C followed by 5 min at 37°C, the final concentration of dNTPs in the extension reaction was 50 μM, and the extension/ligation proceeded for 30–60 min at 37°C. The double-stranded mixture was transformed into JM103, and positive clones were identified by plaque hybridization [16]. Confirmation that the correct change had been introduced was by restriction digestion of the phage RF and by DNA sequence analysis using the upstream synthetic primer illustrated in Figure 2. Single-stranded template from one clone, mp19-5aEco-$S^2$13, was used in a second mutagenesis step designed to introduce a unique BamHI site just 5′ to the start of translation of the *COX5a* gene (Fig. 2). In this step, the synthetic oligonucleotide (5′-CAACTAAGAACGGATCCTACAATGTTACG-3′) introduced two changes to the DNA sequence, a C to G transversion at position −9, and the introduction of an additional C at position −5 (Fig. 2). The procedures used

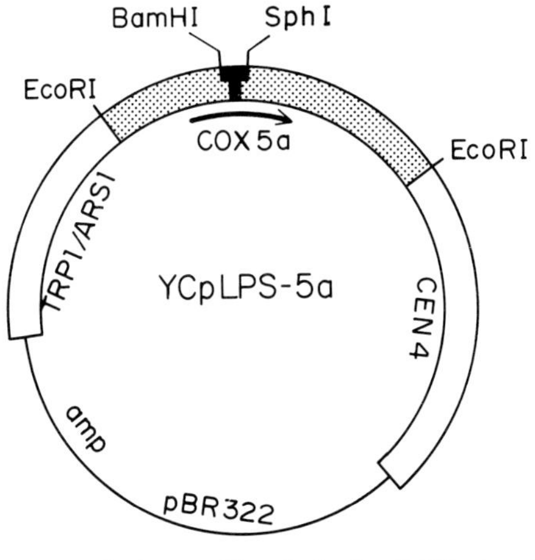

Fig. 1. The plasmid YCpLPS-5a. The leader peptide substitution (LPS) plasmid YCpLPS-5a, a yeast–*E. coli* shuttle vector useful for studies on mitochondrial protein import. Details on the construction of this plasmid are described in Materials and Methods. The BamHI and SphI restriction sites bracket the *COX5a* leader peptide.

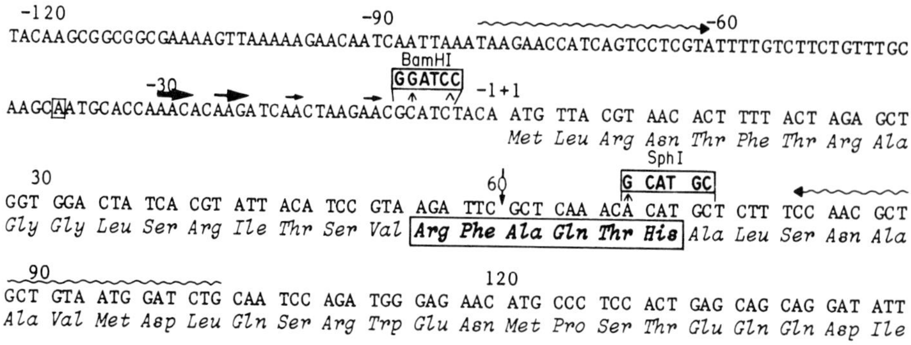

Fig. 2. DNA sequence of the *COX5a* gene in the region of the leader peptide. A partial DNA sequence of the yeast *COX5a* gene [12] from YCpLPS-5a, showing the changes introduced in order to create the two unique restriction sites. The boxed DNA sequences are the result of two separate in vitro mutagenic experiments, using oligonucleotides, as described in Materials and Methods. The boxed A at position −39 has also been changed to a T (see text). The bold arrows represent the 5′ ends of the *COX5a* transcripts; the wavy arrows designate the direction and sequence of two synthetic oligonucleotide primers used in sequencing YCpLPS constructs; and the vertical arrow at position 61 shows the site of processing of the subunit Va precursor [12]. The boxed amino acids from the *COX5a* protein sequence correspond to the boxed arrows from Figure 4. These residues are encoded on the oligonucleotides used to generate the heterologous presequences. Inclusion of the ArgPhe is optional (see text).

were as described above, and, as before, confirmation that the changes were introduced correctly was by restriction analysis and DNA sequencing of replicative form DNA. It should be mentioned that the clone selected for further use, mp19-5aEcoBS^2, which contained both new restriction sites, also carries an additional point mutation introduced by the mutagenesis procedure. However, several experiments, presented in Results, confirm that this mutation, an A to T transversion at position −39 of the *COX5a* 5′ flanking sequence (Fig. 2), has no effect on *COX5a* gene expression.

RF from the phage mp19-5aEcoBS^2 was prepared and cut with EcoRI, and the 3.4-kb fragment containing the *COX5a* region was purified by electroelution. This insert was then ligated into the EcoRI site of the yeast centromere plasmid YCp199, a derivative of YCp19 [17] deleted for the PvuII fragment containing the *URA3* gene (R.M. Wright, personal communication). In the final step, an endogeneous SphI site was deleted from the upstream region of the *COX5a* gene by partial digestion with SphI, purification of full-length linear molecules, treatment with T4 DNA polymerase to flush the ends, followed by religation and transformation into DH1. The final construct, YCpLPS-5a, is shown in Figure 1.

Synthetic Oligonucleotides

Synthetic oligonucleotides used for mutagenesis, as sequencing primers, or for cloning, were synthesized on an Applied Biosystems (University of Colorado, Boulder) or a Biosearch 8600 (University of California, Irvine) DNA synthesizer. In all cases, purification was accomplished by elution from polyacrylamide/urea preparative gels [11]. When the synthetic sequences were to be used to reconstruct leader peptide sequences in YCpLPS-5a, one of two cloning schemes was followed. For a short sequence, in this case LL5a, the oligonucleotide was cloned in single-stranded form by the "notch-cloning" procedure [18]; 50 ng of phosphorylated oligonucleotide was ligated directly into 100 ng of BamHI-SphI cut vector, and the mixture was used to directly transform *E. coli* strain DH1 (Fig. 3). When oligonucleotides coding for longer sequences were to be cloned, a procedure similar to that described by Rossi et al. [19] was followed. Briefly, two strands containing an overlap of about 10 base pairs at their 3′ termini were synthesized separately, annealed, and extended with the Klenow enzyme or reverse transcriptase. After repeated extraction with phenol-chloroform and purification of the DNA by ethanol precipitation, the extended hybrids were cut with BamHI and SphI, ligated into YCpLPS-5a prepared as above, and transformed into DH1 (see Fig. 3). Screening for recombinants produced by either technique was by colony hybridization using an end-labelled oligonucleotide [16]. Before using any recombinant plasmid, it was subjected to DNA sequence analysis to confirm that the correct sequence had been cloned.

DNA Sequence Analysis

In all cases, sequencing was performed by using the dideoxy chain termination method [20]. In order to facilitate the sequence analysis of leader peptide recombinant clones in YCpLPS-5a, a double-stranded method was used. The procedure followed was a modification of that described by Zaug et al. [21]. Routinely, plasmid DNA was denatured in 0.2 N NaOH for 5 min at room temperature and then mixed with 7 ng of a 5′-end-labelled sequencing primer. The mixture was neutralized with NaOAc and ethanol precipitated. Following centrifugation, the pellet was washed, resuspended, and sequenced with AMV reverse transcriptase [11].

Miscellaneous Methods

Transformation of yeast was by the alkali cation [22] method, and transformation of *E. coli* was by the modified method of Hanahan [23]. Established procedures were used for all recombinant DNA and labelling work. Methods for SDS-polyacrylamide gel electrophoresis, immunoblotting, preparation of mitochondria, preparation of nucleic acids, and oxygen consumption assays have been described [11,13].

RESULTS

The Vector YCpLPS-5a

Using the procedures described in Materials and Methods we have constructed the LPS plasmid YCpLPS-5a (Fig. 1). This vector has several important features that make it useful for studies on the import of proteins into mitochondria. First, it is a yeast–*E. coli* shuttle plasmid, containing sequences that allow its propagation and selection in either organism. In addition, when propagated in yeast, the presence of a centromere (*CEN4*) maintains the copy number at approximately one per cell [17]. Also contained on the plasmid is a yeast *COX5a* gene that has been modified by in vitro mutagenesis, so that two unique restriction endonuclease cleavage sites bracket the sequences encoding the Va leader peptide (Fig. 2). When the plasmid is cleaved with these enzymes (BamHI and SphI), this region is removed, and in its place a given test sequence, or "leader peptide cassette," can be substituted. By using synthetic DNA, inserted by either of the two methods shown in Figure 3, virtually any sequence can be positioned in front of subunit Va and assayed for its ability to direct the import of the attached Va polypeptide.

There are several additional features of YCpLPS-5a that are noteworthy. First, during its construction we have not altered the subunit Va polypeptide sequence, nor

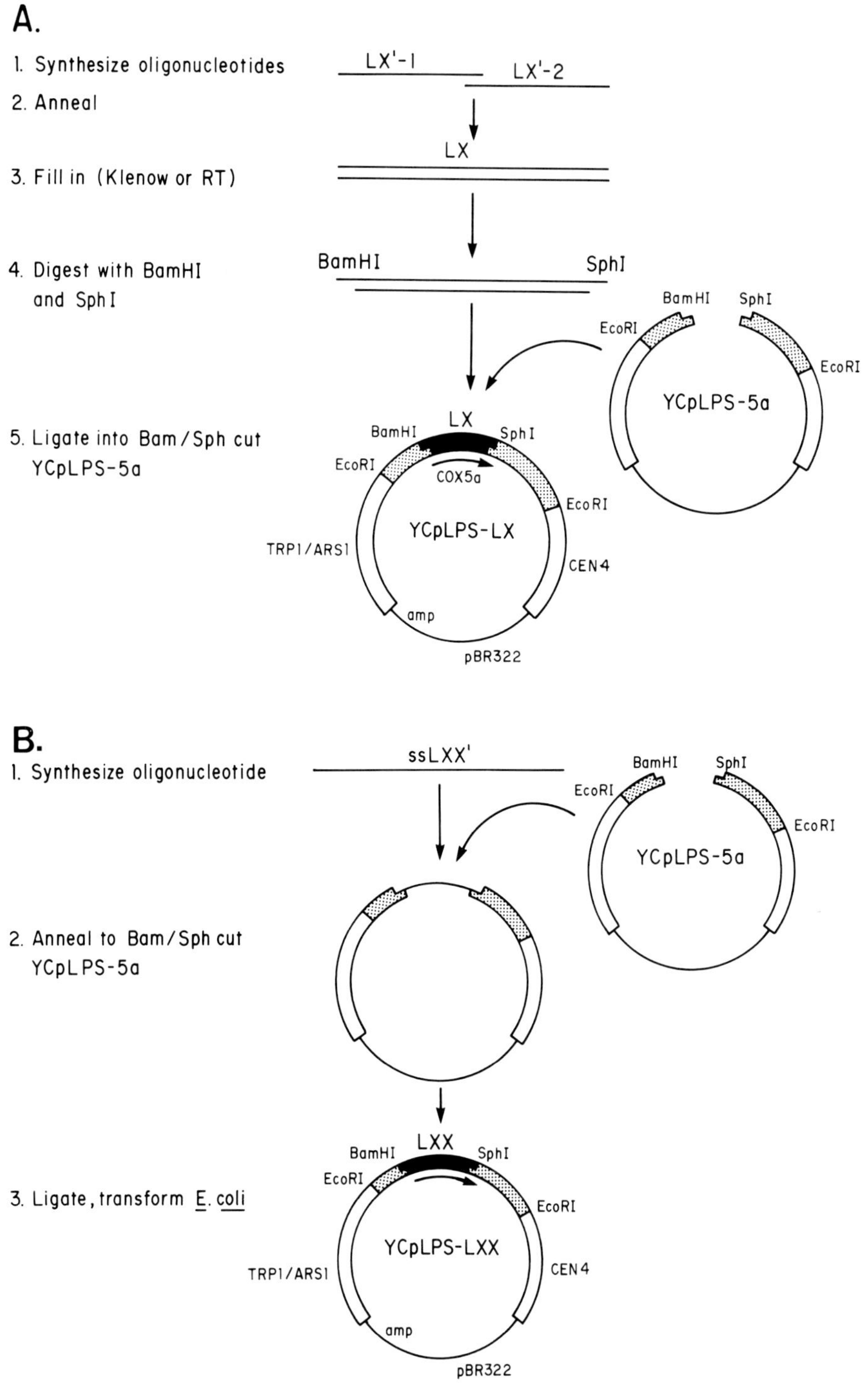

Fig. 3. Construction of YCpLPS derivatives containing heterologous leader peptides. **A:** Schematic representation of the two oligonucleotide method used to generate the hypothetical leader peptide LX (leader X). **B:** The "notch cloning" method used to generate the hypothetical leader LXX (leader XX) [18].

have we changed the translation initiation context (Fig. 2). Second, although creation of the BamHI site introduced three changes to the *COX5a* upstream DNA sequence (Fig. 2), none of the mutations has a detectable effect on the expression of the *COX5a* gene as determined by Northern blot analysis (not shown) and the immunoblot and respiration studies described below.

Another important feature of YCpLPS-5a is the location of the unique SphI site; it lies with the *COX5a* mature sequence at codons 3–5 (Fig. 2). Thus, when synthetic oligonucleotides are used to introduce a given presequence in front of subunit Va, one has the option of retaining or eliminating the amino acid residues immediately before and after the site of proteolytic processing. Therefore, we can begin to examine which residues are important for processing the precursor and whether proteolysis is important for the import or function of subunit Va.

Construction of YCpLPS Derivatives Carrying Heterologous Leader Peptides

In initial experiments, YCpLPS-5a was used to construct several heterologous leader peptide fusions schematically depicted in Figure 4. Using the "notch cloning" procedure first described by Childs et al. [18], and illustrated in Figure 3B for YCpLPS-5a, a *COX5a* gene completely lacking a leader peptide was constructed (YCpLPS-LL5a). It contains only the initiator methionine codon in front of the mature Va sequence. The second derivative shown in Figure 4, positioned sequences coding for the *COX5b* leader peptide in front of subunit Va. As mentioned, the yeast *COX5b* gene is similar to but not identical with *COX5a*. It encodes an isologous polypeptide (Vb) that is functionally equivalent to Va but is not usually observed in mitochondria prepared from aerobically grown, wild-type yeast cells [11–13]. The presequences of these two proteins share 45% homology, significantly less than the 68% homology

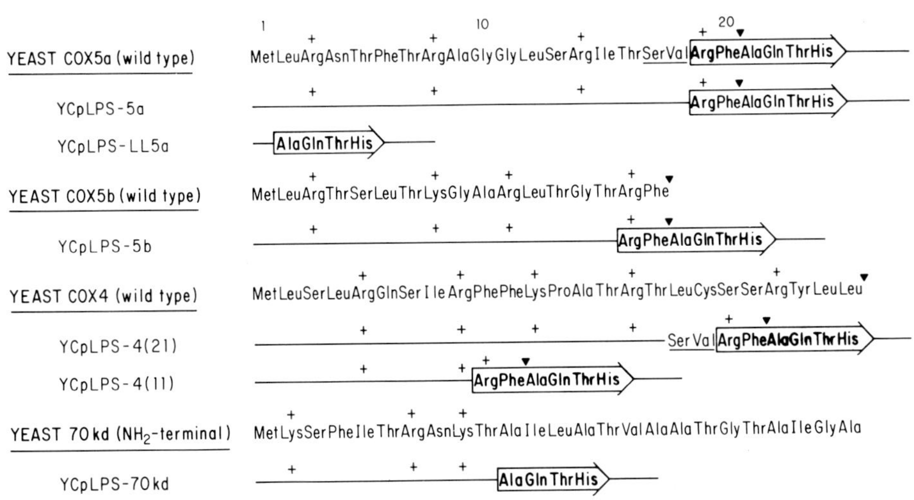

Fig. 4. Constructions in YCpLPS-5a. The polypeptide sequence of the heterologous leader peptides used in this study. The wild-type sequence of the different leader peptides is also shown (we show the amino-terminal sequence for the 70kd protein, since it is not proteolytically processed). Boxed or underlined residues are derived from *COX5a*. Basic residues in the presequence, as well as the *COX5a* proteolytic processing site, are also indicated (+ and arrowhead, respectively).

observed in their respective mature sequences. Substitution of the Vb leader for that of Va permits a comparison of the relative targeting efficiency of each presequence, as well as providing insight into whether the low levels of subunit Vb found in mitochondria prepared from aerobically grown yeast might result from a difference in the inability of this leader peptide to direct the import of its attached subunit [11,13]. The amino acid sequence of the construct, YCpLPS-5b, is shown in Figure 4. It was constructed by using the overlapping oligonucleotide method diagrammed in Figure 3A.

We next constructed two YCpLPS derivatives that carried portions of the yeast *COX4* leader peptide. We chose *COX4* for two reasons. First, it has been used extensively in gene fusion studies on mitochondrial import [1–3,24,25]; consequently, the minimal number of amino-terminal residues required to direct the import of several proteins has been defined [24]. Second, the *COX4* gene product (subunit IV of cytochrome *c* oxidase) appears to be a peripheral membrane protein, located on the inner face of the inner mitochondrial membrane, unlike subunit V, which is imbedded in the bilayer [26–28]. Thus, the use of the subunit IV presequence allows us to determine whether a leader peptide derived from a peripheral inner-membrane-associated protein can correctly localize a heterologous integral inner membrane protein (subunit Va). In the initial subunit IV construct, YCpLPS-4 (21), the first 17 amino acids of the subunit IV leader peptide (previously shown by Hurt et al. to direct mouse dihydrofolate reductase (DHFR) to the mitochondrial matrix) [24] were fused to the COOH-terminal four residues of the wild-type Va leader peptide (Ser-ValArgPhe, Fig. 4). We therefore created a hybrid 21-amino acid leader peptide that contained the amino acids immediately adjacent to the Va processing site. In the second construct, YCpLPS-4 (11), the first nine amino acids of the subunit IV leader peptide (which the same authors also showed could not effectively direct the import of DHFR into mitochondria) [24] were joined to the COOH-terminal 2 (ArgPhe) of subunit Va. As above, the additional amino acids derived from Va were added in order to reconstruct the sequence(s) flanking the wild-type Va processing site. Although these residues increase the length of the leader peptide (by two amino acids) and contribute an extra positive charge as well, they were included because we were interested in determining whether the sequence ArgPheAlaGln constituted a site of proteolytic processing and whether processing was important for the import of subunit Va. This is an interesting question in light of the finding that the NH_2-terminal 12 residues of the *COX4* leader were not processed when used to direct DHFR into yeast mitochondria [24] but that the lack of cleavage did not affect the ability of the protein to be imported. It is also important to mention that the 11 amino acid presequence of YCpLPS-4 [11] differs from the wild-type *COX4* preseqnece at only one position. This is at residue 10, where Arg replaced Phe in the wild-type sequence.

In the last construction shown in Figure 4, YCpLPS-70kd, the subunit Va leader peptide was replaced by the NH_2-terminal 10 residues of the yeast 70-kd protein [29,30]. This construct was made in order to investigate whether a targeting sequence derived from a mitochondrial outer membrane protein (70-kd) can direct an inner membrane protein (subunit Va) to its proper location. While the 70-kd protein does not have a leader peptide that is proteolytically processed, its amino terminus has sequence features reminiscent of the majority of cleaved leader peptides and has been shown to function as a targeting sequence [29,30]. Specifically, it has been demonstrated that the NH_2-terminal 12 residues of this protein direct both subunit IV and

DHFR to the yeast mitochondrial matrix [30]. In YCpLPS-70kd, we have not attempted to reconstruct the processing site from *COX5a* by adding ArgPhe to the presequence.

Import of YCpLPS Derivatives Into Mitochondria

All YCpLPS-5a constructions were transformed into the yeast host JM43-GD5ab (referred to as GD5ab), a strain containing chromosomal disruptions of both endogenous subunit V genes, *COX5a* and *COX5b* [13]. Representative TRP^+ transformants were analyzed for their ability to grow on the nonfermentable substrates glycerol-ethanol and lactate and for staining with the cytochrome oxidase dye TMPD [31]. The results of these experiments are presented in Table I. YCpLPS-5a and its derivatives 5b, 4 (21), 4 (11), and 70 kd complement the subunit V defect in GD5ab and restore positive staining for cytochrome oxidase with TMPD. The control, JM43, also exhibits the predicted result. Table I also indicates that YCpLPS-LL5a, the plasmid encoding the leaderless subunit V gene, fails to complement the defect in GD5ab and stains only very faintly with the dye. To establish that LL5a was expressed in GD5ab, we performed Northern blot and pulse-chase analysis on representative transformants in order to visualize both the LL5a transcript and protein product. The results of these studies (not shown) indicated that the LL5a gene is expressed normally in GD5ab. Since no subunit Va is observed in mitochondria prepared from LL5a transformants (below), we conclude that a leader peptide is required for import of this polypeptide. From the results presented in Table I, we also conclude that targeting sequences derived from proteins located in other mitochondrial compartments can substitute for the Va leader peptide.

Since yeast strains exhibiting significantly less than wild-type levels of cytochrome oxidase activity can grow on nonfermentable substrates (unpublished observations), the results in Table I do not allow us to conclude that the hybrid proteins are targeted as efficiently as wild-type subunit Va. In order to analyze import in a more rigorous manner, we measured the cyanide-sensitive respiration and growth rates (in glycerol-ethanol) of all the strains listed in Table I. The results are shown in Table II. For all strains except those transformed with YCpLPS-LL5a and YCpLPS-70kd, growth and respiration rates are essentially equal to that of JM43. These results, together with the immunoblots shown below, indicate that the 5b, 4 (21), and 4 (11)

TABLE I. Complementation of Growth and Cytochrome *c* Oxidase Deficiency in YCpLPS Transformants

Strain	Complementation[a]		TMPD[b] colony assay
	Glycerol-ethanol	Lactate	
JM43 (no plasmid)	+	+	++++
GD5ab (no plasmid)	−	−	−
GD5ab-YCpLPS-5a	+	+	++++
GD5ab-YCpLPS-LL5a	−	−	±
GD5ab-YCpLPS-5b	+	+	++++
GD5ab-YCpLPS-4(21)	+	+	++++
GD5ab-YCpLPS-4(11)	+	+	++++
GD5ab-YCpLPS-70kd	+	+	++

[a]Complementation refers to growth of the GD5ab transformant on solid lactate media.
[b]The degree of staining in the TMPD colony assay [31] is indicated as ++++ for wild-type activity, ++ for a reduced staining reaction, ± for a faint reaction, and − for no staining reaction.

TABLE II. Growth and Respiration Rates of YCpLPS Transformants

Strain	Cyanide-sensitive respiration[a]	% Wild-type	Growth rate[b]	% Wild-type
JM43 (no plasmid)	33.9	100	2.8	100
GD5ab (no plasmid)	1.0	2.9	na[c]	na
GD5ab-YCpLPS-5a	34.3	101	2.8	100
GD5ab-YCpLPS-LL5a	1.9	5.6	na	na
GD5ab-YCpLPS-5b	38.8	114	2.9	96
GD5ab-YCpLPS-4(21)	43.4	128	2.8	100
GD5ab-YCpLPS-4(11)	nd[d]	nd	3.0	92
GD5ab-YCpLPS-70kd	17.1	50	4	57

[a]Cyanide-sensitive respiration is expressed as pmol O_2 consumed per min OD_{600}. Experiments were performed as described in reference 11 except that the assays were done in 50 mM phosphate buffer, pH 7, containing 0.66% dextrose. Typically, midlog-phase cells, suspended at 1.5 g per 3 ml buffer, were used. Percent wild-type refers to strain JM43 in all cases.
[b]Growth rates were determined in YPEC medium, and are expressed as t_d, the doubling time, in hours. Percent wild-type was calculated by subtracting the percent difference in growth rate of a given strain and JM43 from 100.
[c]na = not applicable.
[d]nd = not determined.

leader peptides function as efficiently as the wild-type subunit Va leader peptide in targeting this subunit to mitochondria.

Although transformants that carry YCpLPS-70kd grow on nonfermentable carbon sources, their growth and respiration rates are significantly lower than wild type (i.e., about half that of JM43 and the other positive transformants). These strains also differ from wild type in other respects. As indicated in Table I, the 70kd construct exhibits reduced staining with TMPD, indicating an intermediate amount of cytochrome oxidase activity. In addition, it takes much longer for YCpLPS-70kd transformant colonies to appear on lactate plates than for the other YCpLPS transformants. Taken together, these results support the conclusion that the 70kd targeting sequence functions less well than the others tested here.

It is interesting that while YCpLPS-LL5a does not restore the ability of GD5ab transformants to respire, it appears to provide a very low level of cytochrome oxidase activity, as determined by TMPD staining (Table I) and cyanide-sensitive respiration (approximately 2% above background, Table II). These results are interesting in that they may indicate an extremely small amount of subunit Va may be capable of entering the mitochondrion without a leader peptide, although this has not been demonstrated directly.

In order to examine the steady-state level of the polypeptide encoded by each YCpLPS construct and to determine if it was proteolytically processed, mitochondria from representative transformants were prepared and examined by immunoblotting, using an antibody to subunit Va. The results of that experiment are presented in Figure 5. It is clear that the steady-state levels of subunit Va observed in transformant mitochondria correspond well with the data from Table I. No subunit Va is observed in GD5ab or GD5ab-YCpLPS-LL5a. Wild-type levels of a mature-sized Va polypeptide can be observed in mitochondria prepared from all other strains, with the exception of GD5ab-YCpLPS-70kd. The 70kd product appears to be larger and less abundant (about 2–3-fold) than wild type. Since the presence of mature-sized subunit Va indicates proteolytic processing of a given preprotein, and since processing

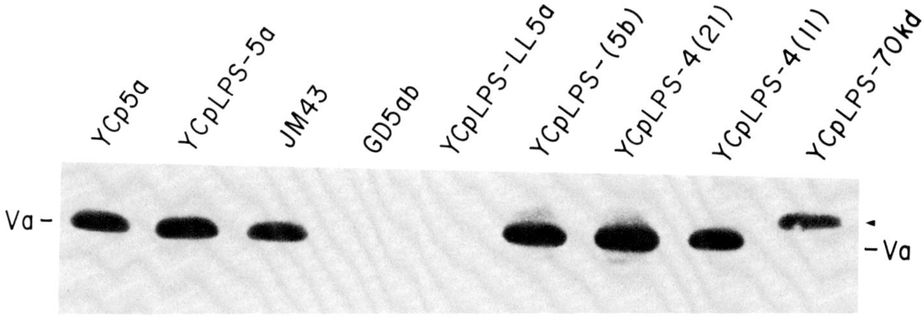

Fig. 5. Steady-state levels of subunit Va in mitochondria. Western immunoblot of mitochondria prepared from the wild-type strain JM43, the *COX5* double null mutant JM43-GD5ab (GD5ab), or GD5ab transformed with the indicated plasmids. In each case, 25 μg of mitochondrial protein was loaded on an 11.8% glycerol-urea polyacrylamide gel and blotted as described [10]. An antisubunit V antisera and ^{125}I-protein A were used to probe the blot. YCp5a is a wild-type *COX5a* centromere plasmid, used as a control [13].

correlates well with the presence of the ArgPheAlaGln sequence (which is absent from YCpLPS-70kd), we suggest that this sequence constitutes at least part of a cleavage site. The finding that the two subunit IV constructs appear to be processed to mature subunit Va is interesting in light of previous results which showed that 16- and 12-amino acid leader peptides derived from subunit IV were not processed when attached to mouse DHFR [24]. Furthermore, cleavage of the YCpLPS-4 (21) leader does not occur solely at an alternative site found in the subunit IV leader (24), since we would have been able to distinguish between the different length mature species in our gel system.

DISCUSSION

In this study we describe a new vector which can be used to conveniently position any sequence at the amino terminus of a well-characterized mitochondrial "passenger" protein. In turn, the ability of that sequence to direct the passenger (the *COX5a* gene product) to the inner mitochondrial membrane can be easily assayed in a phenotypic screen. When this screen (genetic complementation of a strain completely lacking subunit Va) is coupled with additional, more sensitive analyses, an estimate of the efficiency of import can be obtained.

Several of the experimental results presented here are in good agreement with those from other laboratories regarding the targeting of proteins to the mitochondrion. Included among these are that a functional leader peptide is essential for the import of subunit Va and that heterologous leader peptides or targeting sequences derived from proteins located in other mitochondrial compartments can be used to direct the import of subunit Va. However, not all of these heterologous sequences work with equal efficiency.

It is interesting that the 70kd construct functions less well than the others tested here. This result is in line with that of Hurt et al. [30], who found that the NH_2-terminal 12 residues of the 70kd protein also targeted subunit IV to mitochondria with markedly reduced efficiency. Moreover, like YCpLPS-70kd, the *COX4*-70kd construct was not proteolytically processed, yet it is clear that the protein product of either is able to function as a subunit of the cytochrome oxidase holoenzyme. At present, it is not clear if the 70kd constructs function less efficiently because of the extra amino acids at their respective NH_2-termini or because this outer membrane targeting sequence functions less well as an import signal. We favor the latter explanation for two reasons. First, we observe lower steady-state levels of the 70kd-Va polypeptide in transformant mitochondria. Second, we have found that a mutant 32-amino acid leader peptide derived from the human ornithine transcarbamylase gene can effectively target subunit Va to yeast mitochondria. While this leader is not removed from the mature Va sequence during import, GD5ab transformants harboring the construct function normally, as judged by all the criteria used in this study (S.M. Glaser, B.R. Miller, J. Foreman, and M.G. Cumsky, manuscript in preparation).

Two additional conclusions drawn from the results presented here are significant. They are that the passenger protein is probably important for proper targeting, as others have suggested [32], and that the sequence(s) or signal(s) required for correct intramitochondrial sorting of the *COX5a* gene product are contained within the mature Va sequence and not within the leader peptide per se. Evidence in support of the latter conclusion stems from the fact that the NH_2-terminal 11 residues of the subunit IV leader peptide can be used to target subunit Va to the inner membrane with wild-type efficiency. In contrast, the NH_2-terminal 12 amino acids of the same leader direct DHFR to the mitochondrial matrix and subunit IV to the inner face of the inner mitochondrial membrane [24] (A.P.G.M. van Loon, personal communication). Therefore, essentially the same leader peptide targets three different proteins to three distinct intramitochondrial locations, implicating the mature sequence in intramitochondrial sorting. The nature of the subunit Va sorting signal is currently unclear but under investigation.

The experimental system described here should ultimately prove useful for several types of studies on mitochondrial protein import. Here we have used it to test the targeting efficiency of heterologous leader peptides; however, the system should prove useful for additional, broader studies on import, including random mutagenesis of the leader peptide region. Should studies of this nature prove successful in identifying mutants that fail to import the attached protein, revertants that restore import (identified easily by a return to respiratory competancy, i.e., complementation in GD5ab) can then be isolated. In turn, these revertants would prove useful in a genetic analysis of the import pathway. This type of analysis is currently in progress.

ACKNOWLEDGMENTS

The authors gratefully acknowledge the expert technical assistance of Jill Foreman and Lynn Farrell and interesting discussions with Brian Miller. This work was supported by grants from the NIH (GM36675) and CRCC (to M.G.C.) and by grant GM30228 from the NIH (To R.O.P.). S.M.G. is a predoctoral fellow of the UC Biotechnology Training Program.

REFERENCES

1. Douglas M, McCammon MT, Vassarotti A: Microbiol Rev 50:66, 1986.
2. Hurt EC, van Loon APGM: TIBS 11:204, 1986.

3. Hurt EC, Pesold-Hurt B, Schatz G: FEBS Lett 178:306, 1984.
4. Horwich A, Kalousek F, Mellman I, Rosenberg L: EMBO J 4:1129, 1985.
5. van Loon APGM, Brandli A, Schatz G: Cell 44:801, 1986.
6. von Heijne G: EMBO J 5:1335, 1986.
7. Roise D, Horvath S, Tomich JM, Richards JH, Schatz G: EMBO J 5:1327, 1986.
8. Zwizinski C, Schleyer M, Neupert W: J Biol Chem 259:7850, 1984.
9. Ohta S, Schatz G: EMBO J 3:651, 1984.
10. Cumsky MG, McEwen JE, Ko C, Poyton RO: J Biol Chem 258:13418, 1983.
11. Cumsky MG, Ko C, Trueblood CE, Poyton RO: Proc Natl Acad Sci USA 82:2235, 1985.
12. Cumsky MG, Trueblood CE, Ko C, Poyton RO: Mol Cell Biol 7:3511–3519, 1987.
13. Trueblood CE, Poyton RO: Mol Cell Biol 7:3520–3526, 1987.
14. Sherman F, Fink G, Lawrence CW: "Methods in Yeast Genetics." New York: Cold Spring Harbor Laboratory, 1979.
15. Norrander J, Kempe T, Messing J: Gene 23:101, 1983.
16. Zoller MJ, Smith M: DNA 3:479, 1984.
17. Stinchcomb DT, Mann C, Davis RW: J Mol Biol 158:157, 1982.
18. Childs J, Villanueba K, Barrick D, Schneider TD, Stormo GD, Gold L, Leitner M, Caruthers M: In Calender R, Gold L (eds): "Sequence Specificity in Transcription and Translation." UCLA Symposia on Molecular and Cellular Biology, New series, Vol. 30. New York: Alan R. Liss, Inc., 1985, pp 341–350.
19. Rossi JJ, Kierzek R, Huang T, Walker PA, Itakura K: J Biol Chem 257:9226, 1982.
20. Sanger F, Nicklen S, Coulson AR: Proc Natl Acad Sci USA 74:5463, 1977.
21. Zuag AJ, Kent C, Cech TR: Science 224:574, 1984.
22. Ito H, Fukuda Y, Murata K, Kimura A: J Bacteriol 153:163, 1983.
23. Hanahan D: J Mol Biol 166:557, 1983.
24. Hurt EC, Pesold-Hurt B, Suda K, Opplinger W, Schatz G: EMBO J 4:2061, 1985.
25. Allison DS, Schatz G: Proc Natl Acad Sci USA 83:9001, 1986.
26. Cerletti N, Schatz G: J Biol Chem 254:7746, 1979.
27. Power SD, Lochrie MA, Sevarino KA, Patterson TE, Poyton RO: J Biol Chem 259:6564, 1984.
28. Gutweniger H, Bisson R, Montecucco C: Biochim Biophys Acta 635:187, 1981.
29. Hase T, Muller U, Reizman H, Schatz G: EMBO J 3:3157, 1984.
30. Hurt EC, Muller U, Schatz G: EMBO J 4:3509, 1985.
31. McEwen JE, Cameron VL, Poyton RO: J Bacteriol 161:831, 1985.
32. Van Steeg H, Oudshoorn P, Vanhell B, Polman JEM, Grivell LA: EMBO J 5:3643, 1986.

Journal of Cellular Biochemistry 36:289–295 (1988)
Molecular Biology of Intracellular Protein
Sorting and Organelle Assembly 85–91

Uncoupling of Translocation Across Microsomal Membranes From Biosynthesis of Influenza Virus Hemagglutinin

Chuck C.-K. Chao and Phil Bird

Department of Biochemistry, University of Texas Health Science Center at Dallas, Dallas, Texas 75235

This communication presents our recent studies on the biosynthesis of influenza virus hemagglutinin (HA) in a mammalian-cell-free system and its translocation across microsomal membranes. RNAs coding for wild-type (full-length) and mutant (truncated) forms of HA were generated by in vitro transcription by using bacteriophage T7 DNA-dependent RNA polymerase. These RNAs were translated in a rabbit reticulocyte system that was supplemented with dog pancreas membranes, either before translation was initiated or after it had been artificially terminated with the antibiotic cycloheximide. All forms of HA could be cotranslationally translocated. However, only truncated molecules (83% of full length) could translocate after protein synthesis had been terminated. Posttranslational translocation was dependent on the presence of a functional N-terminal signal sequence and occurred only in the presence of ribosomes. The molecular mechanism of protein targeting and translocation across the membrane of the endoplasmic reticulum is discussed based on the signal hypothesis.

Key words: ***in vitro*** **translation, endoplasmic reticulum, influenza hemagglutinin**

For several years, it was believed that translocation of mammalian proteins across the membrane of the rough endoplasmic reticulum (ER) is strictly coupled to their biosynthesis (see [1–4] for reviews). Several components have been identified and characterized that played important roles in this process (eg, signal recognition particle [SRP] [5,6] and SRP receptor, also termed docking protein [7–10]). Using cell-free protein-synthesizing systems derived from wheat germ, it has been shown that SRP specifically arrests the biosynthesis of nascent secretory and transmembrane proteins, presumably by interacting with their amino-terminal hydrophobic signal peptides. Translation resumes after interaction between the ribosome-bound SRP and

Dr. Chuck C.-K. Chao's present address is Department of Medicine/Oncology, Stanford University School of Medicine, Stanford, CA 94305.

Received March 11, 1987; revised and accepted August 5, 1987.

the SRP receptor, an integral membrane protein of the endoplasmic reticulum. The nascent proteins are then transferred across the membrane to the lumenal side of the ER by a process whose molecular mechanisms are not yet understood.

Recent evidence, however, indicates that SRP-mediated translational arrest is less than complete in cell-free systems derived from reticulocyte lysates [11,12]; in some cases, it amounts to no more than a transient delay in biosynthesis of nascent secretory proteins [12–14]. Furthermore, in systems derived entirely from mammalian sources, translocation is not obligatorily cotranslational and can occur after biosynthesis of a protein is complete (see [2–4] for reviews). The list of proteins that can be translocated posttranslationally is growing and includes the human erythrocyte glucose transporter [15], human placental lactogen [16], honeybee prepromelittin [17], and a fusion protein consisting of the signal sequences of β-lactamase and human α-globin [18]. In all these cases, posttranslation translocation is both ATP- and ribosome-dependent. A yeast secretory protein, the precursor to α-factor, can also be posttranslationally translocated in a yeast-cell-free translation system [19–21]. The process is ATP-dependent but, by contrast to the situation in mammalian systems, it occurs in the absence of ribosomes.

To define more precisely the requirements for translocation of proteins in a mammalian-cell-free system, we have analyzed the ability of a well-characterized transmembrane protein, influenza virus hemagglutinin (HA), to be co- and posttranslationally translocated across microsomal membranes. We show that full-length HA can be translocated co- but not posttranslationally. By contrast, truncated forms of the molecule, less than 83% of the wild-type molecule in length, can be translocated across microsomal membranes after their synthesis is completed. This posttranslational translocation, which is comparatively inefficient, requires the presence of ribosomes. A possible pathway for protein translocation, based on these results, is discussed.

MATERIALS AND METHODS

Construction and Mutagenesis of a Vector for Transcription of a cDNA Encoding Influenza Hemagglutinin

Cloned cDNA encoding the HA of the A/Japan/305/57 strain of influenza virus [22] was inserted into an expression vector, pT7/T3-19 (BRL), by using standard methods [23]. To construct a mutation in the N-terminal hydrophobic signal of HA, wild-type HA cDNA was cloned into M13mp18 by using HindIII and BamHI restriction sites. Site-directed mutagenesis of HA cDNA with mismatched oligonucleotides and identification of the desired mutants was carried out as described by Zoller and Smith [24]. Mutant forms of HA cDNA were then subcloned into pT7/T3-19 for in vitro expression. A restriction map of the recombinant vector, designated pT7HA-Jap, and the encoded HA polypeptide are schematically represented in Figure 1.

Translation in Mammalian Cell-Free Systems of Transcripts Generated In Vitro

Transcription of wild-type and mutant forms of HA cDNA in pT7/T3-19 was carried out by using bacteriophage T7 DNA-dependent RNA polymerase according to the manufacturers' instructions (BRL). mRNA was synthesized from 0.5 μg of linearized DNA plasmid and aliquots of the transcription reaction mixture were used

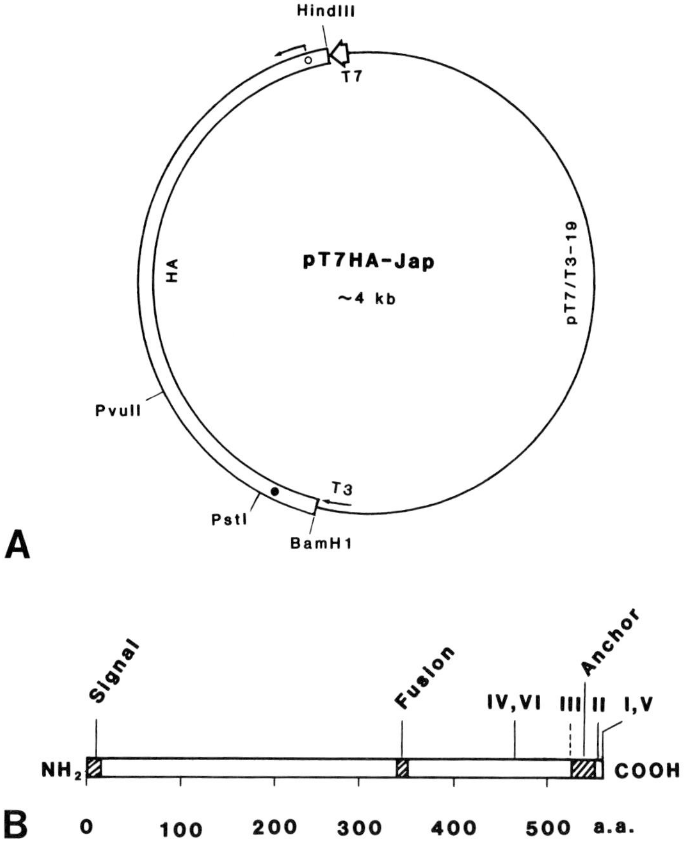

Fig. 1. Restriction map of pT7HA-Jap and schematic representation of HA peptide. **A:** Restriction map of pT7HA-Jap. Only the restriction sites used to linearize HA cDNA are indicated. **B:** Schematic representation of HA polypeptide. Three hydrophobic domains of HA are hatched with their respective functions indicated. Roman numerals represent the full-length (I,V) and truncated forms (II, III, IV, VI) marked with arrows at the end of each molecule. The length scale (amino acids) is also shown below the polypeptide. Restriction sites used to linearize HA cDNA for each polypeptide are: BamH1 for I and V; Pst1 for II; BamH1 for III (the cDNA for this particular case was from independent construct); PvuII for IV and VI.

directly in a cell-free translation system derived from rabbit reticulocytes (Promega Biotech). Each translation reaction contained approximately 0.5 μg of mRNA and was incubated at 30°C for 40 min. Some translation reactions were supplemented with rough microsomes from dog pancreas (a gift from Peter Walter, UCSF) either before or after termination of protein synthesis (see below).

Posttranslational Treatment and Translocation Assays

Cycloheximide (Sigma) was added at the end of translation to a final concentration of 0.1 mg/ml. After storage for 20 min on ice, rough microsomes were then added to some samples and the incubation was continued for a further 40 min. Translocation of HA across microsomal membranes was detected by (1) a decrease in the electrophoretic mobility of the molecule caused by attachment of mannose-rich oligosaccharides and (2) protease-protection assays. In the latter case, duplicate samples were either treated with Triton-X 100 or not treated, and then were digested with proteinase K (0.05 mg/ml) for 20 min on ice. Phenylmethylsulfonyl fluoride (PMSF) (Sigma) was then added to a final concentration of 0.5 mg/ml and the samples were then stored for 15 min on ice. Radiolabeled HA was immunoprecipitated with a polyclonal antiserum that recognizes both native and denatured forms of the protein and then analyzed by SDS–polyacrylamide gel electrophoresis as described previously [25,26]. [^{35}S]methionine-labeled bands were visualized by fluorography.

RESULTS AND DISCUSSION

We have analyzed the translocation of wild-type, truncated, and mutant forms of influenza virus HA into microsomes both during and after completion of biosynthesis in a mammalian-cell-free protein-synthesizing system. Results are summarized in Table I. Typical results are shown in Figure 2. Full-length HA carrying a wild-type signal (panel I) is translocated when rough microsomes are present during biosynthesis of the molecule. The translocated molecules become glycosylated (as judged by a decrease in electrophoretic mobility, lane 2) and resistant to exogenously added protease (lane 5). As expected, disruption of microsomes with Triton-X 100 renders the translocated HA sensitive to protease. Because no translocated HA is detected when microsomes are added to the reaction after synthesis of radiolabeled HA has been terminated by the addition of cycloheximide (lane 4), we conclude that translocation of full-length HA is obligatorily cotranslational in this cell-free system. Cotranslational translocation is dependent on the presence of a functional N-terminal signal sequence, since full-length HA carrying arginine residues in place of the leucines at positions 6 and 8 was not detectably translocated, even when microsomes were present during biosynthesis (panel V, lanes 8 and 9).

By contrast, a truncated form of HA (83% of full-length) was translocated both co- and posttranslationally (panel IV). Although the efficiency varied from experiment to experiment, cotranslational translocation (lane 2) was always much more efficient than posttranslational translocation (lane 3). In the experiment shown in Figure 2, more than 50% of the truncated HA is translocated when microsomes are present during the entire period of incubation; however, less than 5% of HA is translocated when microsomes are added to the system after synthesis of HA has been terminated with cycloheximide. Nevertheless, the small fraction of HA that is translocated into microsomes posttranslationally is both glycosylated and resistant to protease (lanes 5

TABLE I. Summary of the Translocation of Wild-Type and Mutated Forms of HA Across Microsomes

HA construct	Signal	Length, aa[a] %	Translocation Co	Translocation Post
I	WT	562 (100)	+	–
II	WT	555 (99)	+	–
III	WT	536 (95)	+	–
IV	WT	465 (83)	+	+
V	MUT	562 (100)	–	–
VI	MUT	465 (83)	–	–

[a]The number in parentheses represents a relative size of the truncated form to the full-length molecule, which is 562 amino acids (aa) in length.

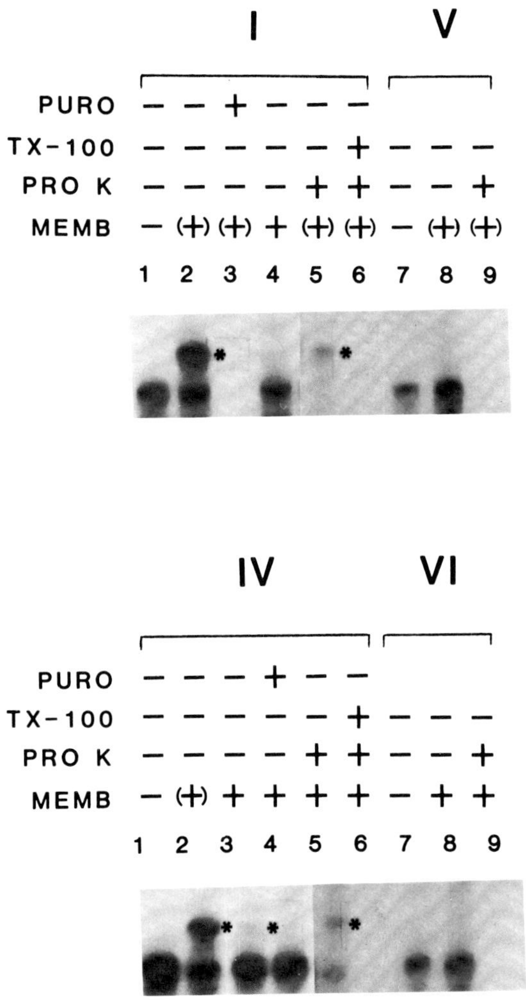

Fig. 2. In vitro expression of HA cDNA. In vitro translation products were immunoprecipitated and analyzed by SDS-PAGE and fluorography. **Upper panel:** Full-length HA, I and V, with wild-type **(lanes 1–6)** or mutated **(lanes 7–9)** signals synthesized either with or without microsomes. Addition of microsomes after translation is marked with "+" symbol; otherwise addition before translation with symbol in parentheses. Lane 3, puromycin and microsomes were included in the reaction mixture before protein synthesis. After translation, the proteins were subjected to the proteinase K treatment in the presence (lane 6) or absence (lanes 5 and 9) of 0.5% Triton X-100. The translocated forms are marked with "*". **Lower panel:** Truncated HA, IV and VI, with wild-type **(lanes 1–6)** or mutated **(lanes 7–9)** signals synthesized in the conditions indicated. After synthesis, proteins were treated as described for the upper panel except that puromycin (lane 4) was added after protein synthesis. Lane 5 was overexposed to visualize a low level of the translocated band (compare lanes 3 and 5). The symbols are the same as for the upper panel.

and 6). Posttranslational translocation of truncated HA was abrogated both by puromycin (lane 4) and by the presence of a mutated signal sequence (compare lanes 3 and 8). We therefore conclude that this process depends on the integrity of the ribosomal-mRNA complex and that it uses the same hydrophobic signal that is responsible for cotranslational translocation.

These results raise several questions. Why should posttranslational translocation be restricted to molecules of HA that are less than full length? Why should the process require a functional signal sequence and integrity of the ribosome-mRNA-nascent protein complex? It seems likely to us that the failure of full-length HA to translocate is a consequence of the proteins' folding into a conformation that is incompatible with its transfer across the microsomal membrane. This hypothesis has been proposed previously by Maher and Singer [27], who suggest that the formation of disulphide bonds in proteins synthesized in cell-free systems is sufficient to prevent their posttranslational translocation. This explanation cannot explain our observations, since the inclusion of reducing agents in the in vitro system did not allow full-length HA to be posttranslationally translocated (unpublished results). Even in the absence of disulfide bond formation, it therefore seems that HA can fold into configuration(s) that are unacceptable for translocation. The longer the polypeptide, the greater are the chances that it will enter such an unacceptable configuration. Full-length HA seems to enter such unacceptable configurations rapidly. However, a truncated form of HA, lacking 97 amino acids from the carboxy terminus, displays some ability to be posttranslationally translocated. Presumably the low efficiency of this process reflects the small proportion of the truncated molecules that have avoided forming structures that cannot be translocated.

The mRNAs that encode truncated forms of HA (Table I) were generated by in vitro transcription of linearized plasmids and therefore do not contain 3′ ends typical of eukaryotic mRNAs. Ribosomes do not dissociate from such artificial mRNAs but remain attached to their 3′ ends as a ribosome-mRNA-nascent polypeptide complex [20]. The integrity of this complex is apparently required for posttranslational translocation. It seems possible that this complex imposes restrictions on the tertiary structure of the nascent polypeptide and therefore serves to maintain it in a conformation acceptable for translocation.

The requirement for a functional signal can be interpreted in two ways: either a stretch of hydrophobic amino acids is required simply to trigger interaction between the polypeptide and the membrane of the microsome, or it is required to interact with signal recognition particle (SRP). In the latter case, SRP may serve to increase the efficiency with which the ribosome complex is attached postsynthetically to the microsomal membrane or it may merely place further constraints on the conformation of the nascent polypeptide. We are therefore currently analyzing the dependence of posttranslational translocation of truncated forms of HA on the presence of SRP.

ACKNOWLEDGMENTS

We thank Joe Sambrook and Mary-Jane Gething for advice, Peter Walter (University of California, San Francisco) for providing dog microsomes, and Carl Sidle for assistance with the figures. This work was paid for by grants from National Institute of Health to J.S. and M.-J.G. P.B. was supported by a postdoctoral fellowship from the Damon Runyon–Walter Winchell Fund.

REFERENCES

1. Walter P, Gilmore R, Blobel G: Cell 38:5, 1984.
2. Wickner WT, Lodish HF: Science 230:400, 1985.
3. Rapoport TA: CRC Crit Rev Biochem 20:73, 1986.
4. Walter P, Lingappa VR: Annu Rev Cell Biol 2:499, 1986.
5. Walter P, Blobel G: Proc Natl Acad Sci USA 77:7112, 1980.
6. Walter P, Blobel G: Nature 299:691, 1982.
7. Walter P, Blobel G: J Cell Biol 91:557, 1981.
8. Gilmore R, Blobel G, Walter P: J Cell Biol 95:463, 1982.
9. Gilmore R, Walter P, Blobel G: J Cell Biol 95:470, 1982.
10. Meyer MI, Krause E, Dobberstein B: Nature 297:647, 1982.
11. Meyer DI: EMBO J 4:2031, 1985.
12. Ibrahimi I: J Cell Biol 104:61, 1987.
13. Lipp J, Dobberstein B, Haeuptle M-T: J Biol Chem 262:1680, 1987.
14. Wiedmann M, Kurzchalia TV, Bielka H, Rapoport TA: J Cell Biol 104:201, 1987.
15. Mueckler M, Lodish HF: Cell 44:629, 1986.
16. Caulfield MP, Duong LT, Rosenblatt M: J Biol Chem 261:10953, 1986.
17. Zimmermann R, Mollay C: J Biol Chem 261:12889, 1986.
18. Perara E, Rothman RE, Lingappa VR: Science 232:348, 1986.
19. Hansen WB, Garcia PD, Walter P: Cell 45:397, 1986.
20. Waters G, Blobel G: J Cell Biol 102:1543, 1986.
21. Rothblatt M, Meyer DI: Cell 44:619, 1986.
22. Gething M-J, Bye J, Skehel J, Waterfield M: Nature 287:301, 1980.
23. Maniatis T, Fritsch E, Sambrook J: "Molecular Cloning." New York: CSHL, 1982.
24. Zoller MJ, Smith M: Methods Enzymol 100:468, 1983.
25. Gething M-J, Sambrook J: Nature 293:620, 1981.
26. Gething M-J, Sambrook J: Nature 300:598, 1982.
27. Maher PA, Singer SJ: Proc Natl Acad Sci USA 83:9001, 1986.

Journal of Cellular Biochemistry 36:323–327 (1988)
Molecular Biology of Intracellular Protein
Sorting and Organelle Assembly 93–97

ADP-ATP Carrier of *Saccharomyces cerevisiae* Contains a Mitochondrial Import Signal Between Amino Acids 72 and 111

Cynthia S. Smagula and Michael G. Douglas

Department of Biochemistry, University of Texas Health Science Center at Dallas, Dallas, Texas 75235

The ADP-ATP carrier (also referred to as the adenine nucleotide translocator) of *Saccharomyces cerevisiae* is encoded by a nuclear gene, translated in the cytosol, and imported into the mitochondrial inner membrane. In order to study the determinants of mitochondrial import, a series of fusion proteins, consisting of the first 21, 72, and 111 amino acids of the ADP-ATP carrier, joined to mouse dihydrofolate reductase were generated. Dihydrofate reductase is a cytosolic protein that does not bind mitochondria. The reticulocyte lysate reaction containing the ^{35}S-methionine-labeled protein was incubated with mitochondria in a buffer containing 3% BSA. Following incubation for import, the reactions were treated with 1 mM PMSF or 25 μg/ml proteinase K; mitochondria were reisolated and analyzed by gel electrophoresis. The 21 and 72 amino acid hybrid proteins showed a low level of binding to mitochondria: the bound form was entirely protease accessible. The 111 amino acid hybrid protein was imported to a protease-protected location within mitochondria. It is concluded that the first 72 amino acids of the ADP-ATP carrier do not suffice to import the protein into mitochondria and that the region between amino acids 72 and 111, a region that contains a transmembrane-spanning domain, constitutes at least part of the mitochondrial import signal.

Key words: intracellular sorting signal, mitochondrial inner membrane protein, in vitro import

Localization of newly translated proteins to intracellular organelles requires a signal in the primary sequence of the protein and a mechanism for recognizing the signal in the organelle. Mitochondrial import signals previously identified have been generally characterized by the presence of a basic N-terminal region, which by itself suffices to direct proteins to the mitochondrial matrix. The presence of a hydrophobic sequence and the distance of the sequence from the matrix-import sequence appear to determine to which of the intramitochondrial compartments the protein localizes [1–3].

Received May 12, 1987; revised and accepted October 13, 1987.

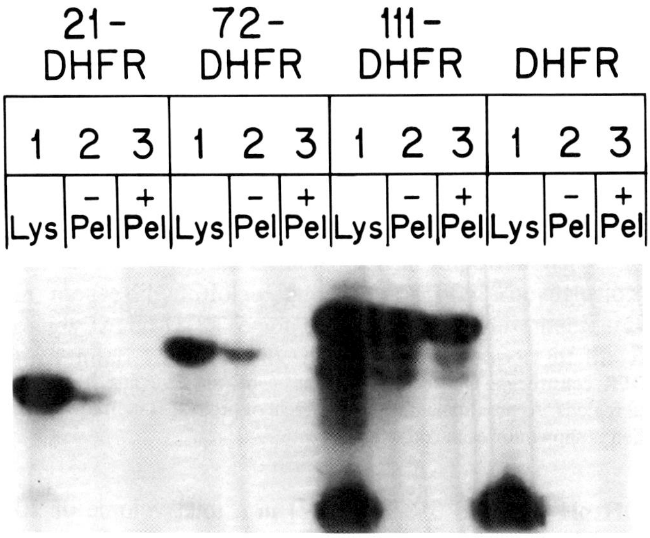

Fig. 2. Binding and import of 21-DHFR, 72-DHFR, 111-DHFR, and DHFR. **Lane 1:** 20% of reticulocyte lysate containing ^{35}S-labeled protein added to import reaction. **Lane 2:** 1 mM PMSF added following import, mitochondria reisolated. **Lane 3:** Import reaction treated with 25 μg/ml proteinase K following import, mitochondria reisolated. Import buffer contained 3% BSA [12].

location in transformed yeast cells [5]. The present results, based on in vitro import of DHFR fusion proteins, indicate the absence of an import signal in the first 72 amino acids of the carrier and the presence of an import signal in the region between amino acids 72 and 111. The absence of an import signal in the extreme N-terminal region suggests that the ADP-ATP carrier is imported by a different mechanism than mitochondrial proteins that contain an import signal in the N-terminal region. Evidence that binding and import of porin and the ADP-ATP carrier of *Neurospora crassa* are blocked by protease treatment of mitochondria under conditions that do not hinder import of ATPase subunits 2 and 9 suggests that import of the carrier and of porin is mediated by a common component that is distinct from components that mediate import of the ATPase subunits [13]. The fact that water-soluble porin competes for the binding site of the ADP-ATP carrier of *Neurospora crassa* [14] reinforces the possibility that the first step of import of the ADP-ATP carrier [12] may be identical with that of porin.

ACKNOWLEDGMENTS

This work is supported by NIH GM 36536 and I-814 from the Robert A. Welch Foundation.

REFERENCES

1. Colman A, Robinson C: Cell 46:321, 1986.
2. Hurt EC, Van Loon APGM: Trends Biochem Sci 11(5):204, 1986.
3. Douglas MG, McCammon M, Vassarotti A: Microbiol Rev 50:166, 1986.
4. O'Malley K, Pratt P, Robertson J, Lilly M, Douglas MG: J Biol Chem 257(4):2097, 1982.

5. Adrian GS, McCammon MT, Montgomery DL, Douglas MG, Mol Cell Biol 6:626, 1986.
6. Simonsen CC, Levinson AD: Proc Natl Sci USA 80:2495, 1983.
7. Chen W-J, Douglas MG: Cell 49:651, 1987.
8. Laemmli UK: Nature 227:680, 1970.
9. Sanger F, Coulson A, Barrell B, Smith A, Roe B: J Mol Biol 143:161, 1980.
10. Maniatis T, Fritsch EF, Sambrook J: "Molecular Cloning, A Laboratory Manual." Cold Spring Harbor, New York: Cold Spring Harbor Laboratory, 1982.
11. Daum G, Bohni PC, Schatz G: J Biol Chem 257:13028, 1982.
12. Pfanner N, Neupert W: J Biol Chem 262:7528, 1987.
13. Zwizinski C, Schleyer M, Neupert W: J Biol Chem 259:7850, 1984.
14. Pfaller R, Neupert W: EMBO J 6:2635, 1987.

Molecular Biology of Intracellular Protein Sorting and Organelle Assembly 99–104 (1988)

Specificity of the Actin Amino Terminal Processing Protease

Peter A. Rubenstein, Larry R. Solomon, and Debra J. Martin

Department of Biochemistry, University of Iowa College of Medicine, Iowa City, Iowa 52242

Our laboratory has been studying the post-translational processing that occurs at the actin amino terminus following release of the completed polypeptide from the ribosome. This processing reaction involves the removal of either an N-acetyl-methionine or N-acetyl-cysteine from the actin amino terminus by a highly specific actin processing protease, which we have described and partially characterized.

In a recent set of experiments, we have studied the processing of class II actins, those whose genes code for a polypeptide beginning with Met-Cys-Asp (Glu) but which actually begin Ac-Asp (Glu). When these actins are made in vitro in the presence of reagents that block acetylation, part of the actin is synthesized with a Met-Cys-Asp (Glu) amino terminus. Under these conditions, our results demonstrate that the actin processing protease can remove both the Met and the Cys residues sequentially, each in an acetylation-dependent fashion.

In another series of experiments, we have begun to study the biological role of the MeHis residue commonly found at position 73. By using oligodeoxynucleotide-directed, site-specific mutagenesis, we have changed this residue to either Arg or to Tyr. For both mutations, no effect was seen on the ability of the actin to undergo amino terminal processing or to bind to DNase I, although involvement of the MeHis had been implicated previously in both cases. The Arg mutation also had no effect on actin polymerization. However, the Tyr mutation did seem to raise the critical concentration of the actin significantly, indicating that for polymerization, the presence of a positive charge at position 73 might be optimal. Studies are continuing, using this technique to determine the specificity sites that allow the processing reaction to occur.

Key words: acetylation, methylation, site-directed mutagenesis

The actin amino terminal region has recently been shown by Sutoh and others to be a potentially important site of interaction between actin and the actin-binding proteins myosin, depactin, and fragmin [1–3]. It is thus an important part of the molecule in terms of understanding actin structure and function. The actin amino terminus is also the site of an unusual sequence of post-translational processing reactions, which, by association, may be important in allowing actin to interact with the proteins described above.

Received February 5, 1987.

Actin genes can be divided into two classes, dictated by the amino termini of the polypeptides for which they code. Class I actins, including *Dictyostelium* [4,5], yeast [6], and beta and gamma avian and mammalian nonmuscle actins [7] are coded for by genes that specify a protein with a Met-Asp (Glu) amino terminus. Class II actins, including those from *Drosophila* [8], sea urchin [9], and avian and mammalian cardiac, skeletal, and smooth muscle actins [7,10] are specified by genes that code for a polypeptide with a Met-X-Asp (Glu), where X is usually Cys. In both cases, the mature protein begins with Ac-Asp (Glu), where Ac is acetyl.

The loss of the initiator Met residue, such as that occurring in the case of a class I actin, is not unusual in eukaryotic proteins. It occurs early in translation, after about 40 amino acids have been polymerized, but usually only if the second amino acid is neutral [11,12]. In the class I actins, the removal of the methionine by the usual methionine aminopeptidase would seem to violate this dictum.

For the class I actins, we have shown both in vivo [13] and in a reticulocyte cell-free system [14,15] that the actin is actually synthesized as a fully translated polypeptide chain containing an Ac-Met-Asp (Glu) amino terminus. This form of actin has a half-life of about 15 minutes in the cell. In a slower, completely post-translational fashion, a specific protease removes Ac-Met, and the new amino terminus is acetylated to yield the mature form of the protein. We call the removal of the acetylated amino acid "processing" and the enzyme responsible the actin-processing protease. If acetylation of the initiator methionine is prevented, then removal of the methionine does not occur.

For the class II actins, under the usual circumstances, the earliest fully translated actin intermediate we detect is one with an Ac-Cys amino terminus [10,16,17]. This observation is consistent with the fact that the penultimate residue, Cys, is small and neutral, allowing the methionine aminopeptidase to work cotranslationally. The result is an amino terminal Cys, which is immediately acetylated. Following completion of translation, Ac-Cys is removed in a processing step, and the new amino terminus is acetylated to yield the mature actin. When a class II actin is made in a reticulocyte lysate under conditions in which protein acetylation is inhibited by 90% [18], at least 75% of the newly synthesized actin is observed without its initiator methionine residue, underlining the fact that for these actins, the Met is removed usually in an acetylation-independent manner. A summary of these two processing pathways is depicted in Figure 1.

We have observed actin-processing activity in a variety of mammalian tissues, including reticulocytes, skeletal muscle, and liver, as well as in the lower eukaryote *Dictyostelium discoideum* [19]. We have demonstrated that a class II actin can faithfully and efficiently be processed by a mouse fibroblast that normally synthesizes only class I actin isoforms [17]. This result suggests that one enzyme removes the acetyl amino acid from both class I and class II actins. An alternative possibility is that two different enzymes are constitutively produced, regardless of whether the cell synthesizes class I or II isoforms. This latter possibility seems unlikely, however, since two peaks of activity are never observed during our enzyme purification procedures. Although we have achieved a partial 400× purification of the enzyme from rat liver, further progress has been slow, due to the increased apparent instability of the enzyme as the enzyme becomes more pure and to the difficulty of the assay, which is based on two-dimensional gel electrophoresis. Currently, efforts are being made to improve the assay procedure by incorporating the use of recombinant DNA techniques.

We do not believe that this actin-processing activity is the same as the enzymes

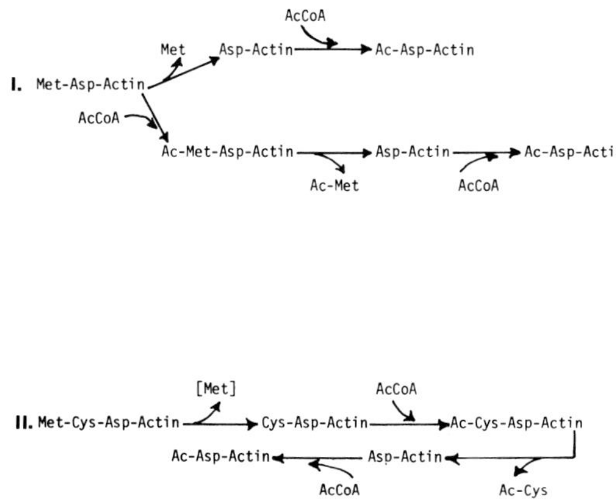

Fig. 1. Amino terminal processing pathways for class I and class II actins. Class I, upper line; pathway that would occur if the initiator Met were removed as a free amino acid early in translation. Class I, lower line; pathway actually followed by class I actins involving post-translational removal of Ac-Met. Class II: pathway usually followed by class II actins. The Met is presumed to be removed cotranslationally as a free amino acid.

described by Brown [20], Manning [21], and Wold [22], which remove an acetyl methionine from the amino terminus of an oligopeptide. Such activities are removed from our preparation during the early stages of purification. It is also unlikely that the phenomenon observed by Lebhertz et al. [23], in which aldolase loses its amino terminal acetyl methionine during isolation of the enzyme from muscle tissue, is caused by the processing enzyme. First, the amino termini of the two enzymes are vastly different. Second, if care is taken during isolation of the enzyme, removal of the Ac-Met from aldolase is prevented, whereas removal of the same residue from actin is complete both in vivo and in vitro soon after translation. Finally, as will be discussed later, the actin-processing protease requires the entire actin molecule in its native form before the reaction will occur.

Our studies so far have indicated some features of the specificity of the actin-processing protease. The enzyme requires the native actin molecule as a substrate, because it will work neither on denatured actin nor on the eighteen-residue amino terminal tryptic fragment of actin. It can accomplish the removal of either Ac-Met or Ac-Cys from the amino terminus of fully translated actin, but it does not have unlimited substrate specificity. *Acanthamoeba* actin is coded for by a gene that prescribes a polypeptide with a Met-Gly-Asp amino terminus [24]. The actin itself contains an amino terminal Ac-Gly-Asp sequence [25]. This is the only actin isolated so far that does not have an acidic residue at its amino terminus.

When *Acanthamoeba* actin was used as a substrate with processing enzyme from either rat liver or *Dictyostelium*, it was completely inert [19]. Furthermore, *Acanthamoeba* itself was devoid of any demonstrable actin-processing activity. Failure to remove the Ac-Gly is not a function of the remainder of the *Acanthamoeba* actin sequence, since this actin is almost identical with *Dictyostelium* actin, which is an excellent substrate for the processing enzyme. Whether any of the other 17 amino acids can be removed in acetylated form by the enzyme is unknown at this time.

If the initiator methionine of class II actins is removed as a free amino acid early in translation by the methionine aminopeptidase, its removal should not be affected by

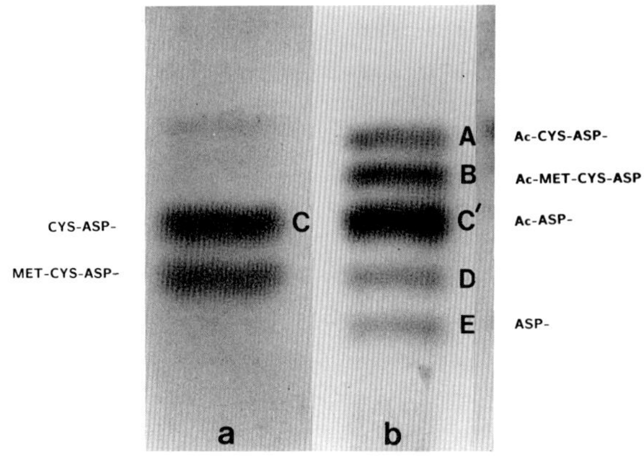

Fig. 2. Processing of *Drosophila* actin in vitro when synthesized in the presence of an acetylation inhibition system. ^{35}S-Met-labeled actin, synthesized in vitro in a reticulocyte lysate in the presence of reagents that inhibit acetylation (lane a), was combined with processing enzyme and acetyl CoA, and the reaction mixture was incubated for 60 minutes (lane b). Shown is an autoradiogram of the amino terminal tryptic peptides of the isolated actins, with their identity labeled [18].

reagents that inhibit protein acetylation. To test this hypothesis, we translated *Drosophila* actin, a class II species, in a reticulocyte lysate under conditions in which acetylation was blocked [18]. Following amino terminal analysis of the resulting actins that were synthesized, 75% were missing the initiator methionine, indicating that the majority of the methionine was removed in an acetylation-independent fashion. However, the remaining 25% contained a Met-Cys-Asp- amino terminus. When acetyl CoA and fresh processing enzyme were added back to these actin intermediates, five species, shown in Figure 2, could be identified. When the time course of conversion of the precursor to the product actins was determined by quantitating the appearance and disappearance of these amino terminal tryptic fragments following isolation of the actins, the methionine-containing species was first acetylated, and the methionine was subsequently removed. If acetyl CoA was omitted, the methionine remained intact. Our results, summarized in Figure 3, thus indicate that while most of the time the ribosomal methionine aminopeptidase removes the initiator methionine of a class II actin as a free amino acid, the actin-processing protease can remove both the initiator Met and the subsequent Cys residues in an acetylation-dependent fashion if given the opportunity. The ability of this enzyme to sequentially remove multiple amino acids from the amino terminus of actin precursors may be very important in the case of Naegleria actin, if the processing enzyme in this organism is like those we have studied to date. This actin is synthesized with four extra amino acids, Met-Ser-Phe-Cys- before the residue that will be the ultimate amino terminus [26].

A second modification that is common to most actins is a 3-MeHis residue at position 73. Although the function of this modified residue is currently unknown, earlier work by Saborio and Palmer [27] led them to suggest that this unusual amino acid might be important in the amino terminal processing of actin. When actin is processed in the presence of reagents that inhibit its reacetylation, it undergoes a change in isoelectric point that can be detected on isoelectric focusing gels. When Saborio carried out an actin translation experiment in vitro in the presence of a mixture of inhibitors of S-adenosylmethionine (SAM)-dependent methylation reactions, the actin isoelectric point

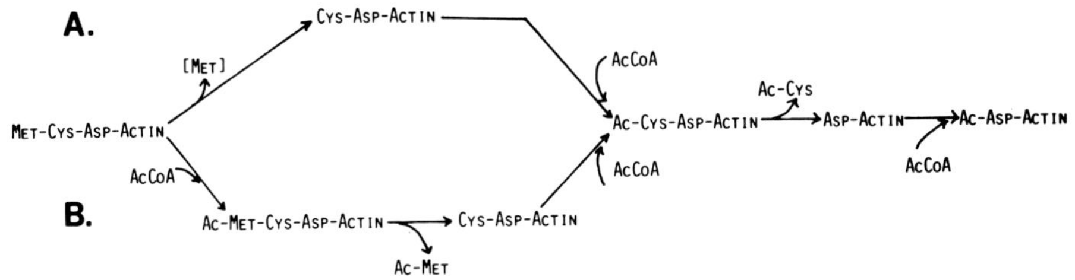

Fig. 3. An alternative pathway for class II actin processing when the initiator Met is not removed early in translation as a free amino acid (B). The normal pathway is shown in A.

shift was inhibited [27]. Since the only known SAM-dependent step in the biosynthesis of actin is the methylation of His_{73}, they suggested the interpretation stated above.

In studying this problem further, we discovered, using a two-stage translation-processing system, that if the SAM inhibitors were present during actin biosynthesis, its subsequent processing was inhibited 50%. If, however, the inhibitors were introduced after completion of translation but during the processing reaction, they produced no significant inhibition of the reaction.

To explore the role of the 3-MeHis further, we decided to eliminate His_{73} from the sequence entirely, using the techniques of site-directed mutagenesis, expressing the mutant actins in a COS-1 cell transient expression system and analyzing the resulting actins to determine if they were processed or not. For this study, we replaced His_{73} with either Arg, which maintained the positive charge of the MeHis, or with Tyr, which contained a neutral hydrophilic phenolic group. After analyzing the amino terminal tryptic peptides of the two mutant actins, the results clearly indicated that both altered proteins were fully processed at the amino terminus. Thus the SAM-dependent step in processing identified by Saborio and by us is not involved with histidine methylation. These same actin mutants were also tested for their ability to bind to DNAse I and to polymerize into F-actin.

Previous results had also indicated that the MeHis might be involved with these processes as well. The MeHis lies very near an interaction site between actin and DNAse I [28], and Naegleria actin, the only actin described so far without a 3-MeHis residue, binds to DNAse I with a greatly reduced affinity in comparison with other actins [29]. Nevertheless, neither of these mutations had any demonstrable effect on the ability of the actin to bind to DNAse I. The His to Arg mutant had no observable effect on polymerization. However, for the His to Tyr mutation, the critical concentration for polymerization was significantly higher than that of the wild-type molecule. It is possible that a positive charge at this position permits the optimal interaction between two actin monomers in the formation of F-actin. That this region of the protein might be important in polymerization has been suggested by Collins and Elzinga [30], based on earlier work [31], which showed that nitration of the neighboring Tyr_{69} prevented F-actin formation.

We have also converted Asp_{11} to either a Asn or a Glu. Asp_{11} is part of a conserved pentapeptide block, Asp-Asn-Gly-Ser-Gly, and the Asp has been identified as a possible binding site for myosin. When we carried out an analysis of the Asp-to-Glu mutation, partial effects were seen on either processing or on DNAse I binding. However, conversion of the Asp to Asn drastically inhibited both of these functions. On the other hand, polymerizability of this latter mutant actin was at least comparable to that of wild-type actin, showing that the mutation did not totally denature the protein. These results identify Asp_{11} either as a recognition site for the processing protease or indicate that it is an important residue for controlling the conformation of the amino terminal end of the molecule. Further studies of this region, which we hope will elucidate the role of amino terminal processing, and of the amino terminal tryptic peptide region in actin function are currently in progress.

ACKNOWLEDGMENTS

This work was supported by a grant from the National Institutes of Health GM-33689 to P.A.R.

REFERENCES

1. Sutoh K: Biochemistry 21:3654–3661, 1982.
2. Sutoh K, Mabuchi I: Biochemistry 23:6757–6761, 1984.
3. Sutoh K, Hatano S: Biochemistry 25:435–440, 1986.
4. Firtel RA, Timm R, Kimmel AR, McKeown M: Proc Natl Acad Sci USA 76:6206–6210, 1979.
5. Vandekerckhove J, Weber KK: Nature 284:475–477, 1980.
6. Ng R, Abelson J: Proc Natl Acad Sci USA 77:3912–3916, 1980.
7. Gunning P, Ponte P, Okayama H, Engel J, Blau H, Kedes L: Mol Cell Biol 3:787–795, 1983.
8. Fyrberg EA, Bond BJ, Hershey ND, Mixter KS, Davidson N: Cell 24:107–116, 1981.
9. Cooper AD, Crain WR: Nucleic Acids Res 10:4081–4092, 1982.
10. Strauch AR, Rubenstein PA: J Biol Chem 259:7224–7229, 1984.
11. Burstein Y, Schechter I: Biochemistry 17:2392–2400, 1978.
12. Tsunasawa S, Stewart JW, Sherman F: J Biol Chem 260:5382–5391, 1985.
13. Rubenstein PA, Martin DJ: J Biol Chem 258:3961–3966, 1983.
14. Rubenstein PA, Smith P, Deuchler J, Redman K: J Biol Chem 256:8149–8155, 1981.
15. Redman K, Rubenstein PA: J Biol Chem 256:13226–13229, 1981.
16. Rubenstein PA, Martin DJ: J Biol Chem 258:11354–11360, 1983.
17. Solomon LR, Rubenstein PA: J Biol Chem 260:7659–7664, 1985.
18. Martin DJ, Rubenstein PA: J Biol Chem 262:6350–6356, 1987.
19. Redman KL, Martin DJ, Korn ED, Rubenstein PA: J Biol Chem 260:14857–14861, 1985.
20. Gade W, Brown JL: J Biol Chem 253:5012–5018, 1978.
21. Jones WM, Manning LR, Manning JM: Biochem Biophys Res Commun 139:244–250, 1986.
22. Radhakrishna G, Wold F: J Biol Chem 261:9572–9575, 1986.
23. Lebhertz HG, Bates OJ, Bradshaw RA: J Biol Chem 259:1132–1135, 1984.
24. Nellen W, Gallwitz D: J Mol Biol 159:1–18, 1982.
25. Vandekerckhove J, Lal AA, Korn ED: J Mol Biol 172:141–147, 1984.
26. Remillard SP, Lai EY, Fulton C: J Cell Biol 99:440a, 1984.
27. Saborio J, Palmer E: J Neurochem 36:1659–1669, 1981.
28. Sutoh K: Biochemistry 23:1942–1946, 1984.
29. Sussman DJ, Sellers JR, Flicker P, Lai EY, Cannon LE, Szent-Gyorgi AG, Fulton C: J Biol Chem 259:7349–7354, 1984.
30. Collins J, Elzinga M: J Biol Chem 250:5915–5920, 1975.
31. Elzinga M: Abstracts of the American Chemical Society Meeting, Washington D.C., 61, 1971.

Molecular Biology of Intracellular Protein
Sorting and Organelle Assembly 105–108 (1988)

Discussion Summary: Genetic Techniques for Studying Intraorganelle Transport

Marjorie C. Brandriss and Lee McAlister-Henn

Department of Microbiology and Molecular Genetics, University of Medicine and Dentistry of New Jersey—New Jersey Medical School, Newark, New Jersey 07103 (M.C.B.); and Department of Biological Chemistry, California College of Medicine, University of California, Irvine, California 92717

Genetic techniques have been used with great success in recent years to: 1) elucidate targeting information in proteins translated on cytoplasmic ribosomes and localized to different organelles; and 2) isolate mutants with defects in the cellular apparatus required for specific localization. Such mutants provide an approach for determining the number of individual components in a localization pathway and for subsequent definition of the biochemical contribution of each component. The first part of this workshop focused on current genetic approaches for the isolation of *Saccharomyces cerevisiae* mutants with defects in the pathway(s) for protein localization to mitochondria. Participants described a number of different selection schemes designed to obtain localization mutants as well as the types of mutants so obtained. One critical question concerned whether current selection criteria, in particular conditional lethality and the accumulation of precursor forms of mitochondrial proteins, are appropriate for isolation of mutants with defects in all steps of the localization pathway. As described with specific examples below and in recent reviews [1,2], several of the localization mutations that have been isolated in *S. cerevisiae* have deleterious effects on cell viability. The reasons for this are unclear, since yeast can grow in the absence of oxidative metabolism. The latter part of the workshop focused on genetic analyses of targeting information contained within mitochondrial proteins encoded by nuclear genes. Key questions concerned the nature of information specifying import and/or processing of precursors and the suborganellar localization of mitochondrial proteins in yeast.

MUTATIONS IN THE LOCALIZATION PATHWAY

The *mas1* and *mas2* (mitochondrial assembly) mutations of yeast were isolated from a collection of 1,200 temperature-sensitive mutants by screening for the accumulation of precursor forms of mitochondrial proteins at restrictive temperatures [3]. An additional screen of 2,000 temperature-sensitive mutants and complementation tests with

Received March 5, 1987.

the Hartwell/McLaughlin collection of temperature-sensitive mutants provided new alleles of *mas1* and *mas2* but no new mutations. Mitochondria isolated from cells containing a *mas1* mutation exhibited pleiotropic effects with respect to processing of imported mitochondrial proteins in vitro. The mature form of the *MAS1* gene product is a mitochondrial matrix protein with a molecular weight of 48,000 and is derived from a 51,000-molecular weight precursor polypeptide. The protein in dimeric form is believed to be the matrix-localized protease responsible for processing mitochondrial signal sequences [4]. The *MAS2* gene product [5] is a polypeptide with a molecular weight of 44,700. Preliminary evidence suggests the protein is localized in the cytoplasm but binds mitochondria in vitro. The extent of defective precursor protein import in vitro varies for mitochondria isolated from strains containing different alleles of *mas2*. The *MAS2* gene product is essential for cell viability.

The inability of yeast strains carrying an F_1 ATPase β-subunit-β-galactosidase hybrid protein to grow on glycerol was used as a selection scheme for the isolation of mitochondrial import-defective mutants [6]. The hybrid protein had previously been shown to be delivered to the mitochondrial inner membrane [7], although the precise cause of the strain's respiratory deficiency is not understood. The *mad1* (mitochondrial assembly and delivery) mutant is a glycerol-utilizing revertant that is temperature-sensitive for growth on all media, accumulates precursor forms of the β-subunit as well as other mitochondrial proteins, and has slightly altered intramitochondrial localization of the hybrid protein. The hybrid β-glalactosidase is still mitochondrial, but its location is now apparently outside the mitochondrial inner membrane.

The *MAD1* gene is essential for viability, and the DNA contains an open reading frame encoding a polypeptide with a molecular weight of approximately 58,000, lacking any obvious mitochondrial targeting sequence. The *MAD1* gene is not allelic with either *MAS1* or *MAS2*, and its gene product is estimated to be approximately 30% as abundant as the β-subunit of F_1 ATPase.

A new approach for isolation of mutants in mitochondrial import involves a competition between nuclear and mitochondrial localization signals [8]. A hybrid protein between the nuclear regulatory protein *HAP2* and β-galactosidase was shown to have both *HAP2* function and its nuclear localization [9]. DNA encoding the mitochondrial targeting signal from the *HEM8* gene, whose product is the mitochondrial matrix enzyme ferrochelatase, was fused to the 5′ end of the *HAP2-lacZ* construct. This *HEM8-HAP2-lacZ* gene fusion introduced into yeast cells produced a tripartite protein that was now mitochondrial and, as a consequence, failed to complement the *hap2* mutation. The resulting respiratory deficient strain was used in a selection for revertants that could grow on nonfermentable carbon sources. The revertants fall into both *cis* (plasmid-borne) and *trans* classes that are being sorted into various complementation groups.

One general suggestion of interest that arose during these discussions was that determination of the abundance of gene products identified through mutations in the mitochondrial localization pathway might aid in distinguishing functions, for example, a soluble carrier vs. a receptor or a processing enzyme.

SIGNAL SEQUENCE ANALYSES

Import information within the amino terminal region of the precursor for the β-subunit of F_1 ATPase (encoded by *ATP2*) has been examined by construction of gene deletions in vitro to remove regions containing basic amino acid residues currently thought to play a role in mitochondrial import [10]. Two such regions within the leader presequence and one near the amino terminus of the mature protein can be removed independently or in pairwise combinations without dramatic impairment of import in vivo, suggesting that import information may be redundant. Simultaneous deletion of all three regions produces a glycerol$^-$ phenotype and very low levels of a cytoplasmically localized β-subunit. Deletion of the signal sequence region from the amino terminus of the mature protein blocks processing but not localization or function of the β-subunit in vivo. The variance of these results with those of in vitro experiments using isolated mitochondria, which show that precursor proteins containing various deletions were not imported or processed, prompted a discussion of the possibility that sufficient nonspecific import could occur in vivo to provide complementation of the glycerol$^-$ phenotype or that the in vitro system does not closely approximate the in vivo import.

Targeting information for the mitochondrial isozyme of alcohol dehydrogenase (encoded by *ADH3*) has been examined through deletion and point mutation analyses of the presequence [11]. Mutant enzymes containing 14 amino terminal residues of the 28 amino acids in the presequence are targeted and imported into the matrix, whereas enzymes lacking this region can associate with the mitochondrial membrane but are not imported. Mutant enzymes containing a variety of amino acid changes in the presequence are correctly targeted and imported, although the rate of processing in vivo is decreased by alterations within the carboxy terminal regions of the presequence, suggesting that targeting and processing information may be distinct. The precursor form of the *ADH3* protein forms a tetrameric structure in vitro analogous to the mature protein structure. Discussion revealed similar results for other mitochondrial precursors that have been examined. However, the relevance of quaternary conformation to the import process in vivo is not clear.

Information required for targeting of an inner membrane protein is being investigated through fusions of different presequences with the mature form of the Va subunit of cytochrome c oxidase (encoded by *COX5a*) [12]. A fusion with an 11-amino acid presequence from subunit IV (encoded by *COX4*), a peripheral membrane protein, permits complementation of a *cox5a* mutation, suggesting that determinants for inner membrane localization may be inherent in the mature protein. Various mutant forms of the human ornithine transcarbamylase leader sequence [13], when fused to the Va subunit, also provide complementation in vivo but are not processed. In vitro analyses are in progress. It was noted that in vivo complementation tests for mitochondrial mutants should be assessed in light of the relatively free accessibility to many metabolic intermediates that may mask the exact suborganellar localization of some enzymes. In addition, the amount of protein that needs to be localized to the mitochondrion may be very small and still allow complementation.

REFERENCES

1. Douglas MG, McCammon MT, Vassarotti A: Microbiol Rev 50:166, 1986.
2. Brandriss MC: Das, Robbins (eds): "Protein Transfer and Organelle Biogenesis." New York: Academic Press, in press, 1988.
3. Yaffee M, Schatz G: Proc Natl Acad Sci USA 81:4819, 1984.
4. Yaffee M, Ohta S, Schatz G: EMBO J 4:2069, 1985.
5. Jensen R, Suda K, Schatz G, Schatz M: J Cell Biochem [Suppl] 11A (Abstract):289, 1987.
6. McCammon MT, Douglas MG: J Cell Biochem [Suppl] 11A (Abstract):280, 1987.
7. Douglas MG, Geller B, Emr S: Proc Natl Acad Sci USA 81:3983, 1984.

8. Greene JR, Guarente LP: Unpublished abstract.
9. Pinkham JL, Olesen JT, Guarente LP: Mol Cell Biol 7:578, 1987.
10. Bedwell DM, Emr SD: J Cell Biochem [Suppl] 11A (Abstract):270, 1987.
11. Pilgrim D, Young ET: Mol Cell Biol 7:294, 1987.
12. Glaser SM, Cumsky MG: J Cell Biochem [Suppl] 11A (Abstract):388, 1987.
13. Horwich A, Kalousek F, Fenton W, Pollock R, Rosenberg L: Cell 44:451, 1986.

Molecular Biology of Intracellular Protein
Sorting and Organelle Assembly 109–116 (1988)

Specificity Determinants for N-Terminal Acetylation and Removal of N-Terminal Acetylmethionine

Gudepu Radhakrishna, Jeffrey Augen, and Finn Wold

Department of Biochemistry and Molecular Biology, University of Texas Medical School at Houston, Houston, Texas 77225

We have found an enzyme in rabbit muscle that catalyzes the specific removal of acetylmethionine from both short (3 residues) and long (11 residues) acetylated peptides. We have not found it to remove any other acetylamino acids from peptides, and this enzyme could thus play a role in the specific removal of the initiator methionine after an obligatory acetylation step. The relative importance of this acetylmethionine-cleaving enzyme and the more generally considered methionine aminopeptidase cannot be assessed at this time. Scoring the individual residues for their statistical appearance in proteins as secondary structure (helix, sheet, turn) formers (+), indifferent (0), or breakers (−) and then comparing the first 40 residues of a large number of acetylated and nonacetylated eukaryotic proteins from the data bank, we found that an Apple IIc computer could separate the acetylated from the nonacetylated ones using the "perceptron algorithm" developed by artificial intelligence researchers some 30 years ago. The major areas of distinguishing structural features appear to be in the sequences 1–10, 16–24, 34–40, and on the basis that the separation of the two groups is unsuccessful if less than 40 residues are used for the comparison, it appears that all three regions of information are involved in the process. It should consequently be reasonable at this time to explore the possibility that both the removal of the initiator methionine and the subsequent acetylation of the newly exposed residue are regulated by the secondary structure features encoded into the first 40–50 residues.

Key words: Ac-aminoacyl-peptide hydrolase, N-terminal processing

The current status of our knowledge of N-terminal processing in prokaryotes and eukaryotes can perhaps be summarized as follows. For prokaryotes, the situation is relatively simple. The gene specifies a Met initiator, and the synthetic apparatus requires that the initiator Met be formylated prior to the start of protein synthesis [1,2]. The isolated in vivo products of protein synthesis rarely, if ever, retain the formyl-Met N-terminus; the resulting unblocked N-terminals do not appear to be a random distribution of amino acids, in that Met, Ala, and Ser make up more than 95% of the observed

Received March 9, 1987.

N-terminal residues [3]. Thus the minimum N-terminal processing for proteins is removal of formate and then, in about 50% of the proteins, removal of Met. Proteins destined for export or membrane residence may have much more extensive trimming of N-terminal signal sequences, but for cytosolic proteins these minimum processing requirements can be satisfied by the N^{α}formyl-protein hydrolase established in several organisms in the 1960s [4], and by a specific Met-aminopeptidase that has recently been found in *E. coli* [5] and *Salmonella* [6].

For the eukaryotic organisms (Fig. 1), the situation is complicated by the fact that the initiator methionine is removed for a majority of the known sequences and that the N-terminal residue may be acetylated. [A number of other more rarely encountered and specialized N-terminal derivatives are also known (7) but will not be considered here; some of them are discussed elsewhere in this volume.] Eliminating all proteins for which N-terminal signal sequences are removed, but including one secreted protein (ovalbumin), which may acquire its acetyl-group during passage through the secretory pathway, the illustrations in Figure 1 clearly document that for several proteins the N-terminal modification is limited to acetylation of Met or the removal of the initiator methionine, followed in some cases by acetylation of residue 2. In one well-documented case (actin), discussed by Rubenstein in the Met is removed as Ac-Met-. The reactions to account for all these observed facts thus must include all the enzymes given in the accompanying scheme in Figure 1 [8–15]: step 1, Met-aminopeptidase [as discussed by Bradshaw, 17] identical to or distinct from other aminopeptidases (step 2); step 3, Methionyl-protein-transacetylase identical to or distinct from other transacetylases (step 4); step 5, AcMet-peptide hydrolase, identical to or distinct from other Ac-amino acyl-peptide hydrolases (step 6). Enzyme 2, the nonspecific aminopeptidase, is the only one for which there is no documented involvement in biosynthetic N-terminal processing; all the others are required either cotranslationally or post-translationally to explain the available data, and all of them—transacetylases [16], Met aminopeptidase [17], and a fairly broad spectrum of acetylamino acyl-peptide hydrolases [18–20], including the specific AcMet-peptide hydrolase that is one of the topics of this discussion [21]—indeed appear to have been identified in at least one eukaryotic source.

We reported last year that rabbit muscle contains an enzyme that appears to cleave AcMet from the N-terminus of peptides [21]. The enzyme was originally observed in the ribosomal (and microsomal) fractions but could be liberated from the ribosomes by high salt extractions. The enzyme has now been purified about 1,400-fold to sufficient purity to assess some of its properties. Based on gel filtration and SDS gel electrophoresis, it is a single polypeptide chain with an M_r of about 85,000 (see Note Added in Proof). Is it a "ribosomal protein"? If our definition of a ribosomal protein is limited to the approximately 80 characterized proteins found reproducibly after dissociation of eukaryotic ribosomes [22], and presumably existing in the original complex in stoichiometric amounts, the enzyme does not qualify. The largest protein found in ribosomes has a molecular weight of about 50,000, so the enzyme is too large. In addition, based on the specific activity of the enzyme in the ribosomal pellet [21] compared with that of the pure enzyme, its concentration in the ribosomes could only have been about 0.2 mol/mol of ribosomes. Thus the exclusive presence of the enzyme in the ribosomal and microsomal fractions of the muscle homogenate may merely represent an artifact of isolation; if the enzyme is indeed associated with ribosomes in vivo, it must be a transient, nonstoichiometric association, comparable perhaps with the interaction of initiation and elongation factors with the ribosomes at certain stages of their activity.

A. N-terminal Structures

Troponin C (Avian slow muscle)	cDNA:	Met-Asp-Asp-Ile-Tyr- (8)
	Protein:	Ac-Met-Asp-Asp-Ile-Tyr-
Aldolase (Muscle)	cDNA:	Met-Pro-His-Ser-His- (9)
	Protein:	Pro-His-Ser-His- (10)
		(Ac-Met-Pro-His-Ser-His-)
Enolase (Yeast)	cDNA:	Met-Ala-Val-Ser-Lys- (11)
	Protein:	Ala-Val-Ser-Lys- (12)
Actin-type I (Slime mold)	cDNA:	Met-Asp-Asp-Asp-Ile- (13)
	Protein:	Ac-Asp-Asp-Asp-Ile-
Actin-type II (Muscle)	cDNA:	Met-Cys-Asp-Glu-Asp- (13)
	Protein:	Ac-Asp-Glu-Asp-
Ovalbumin (Avian eggs)	cDNA:	Met-Gly-Ser-Ile-Gly- (14)
	Protein:	Ac-Gly-Ser-Ile-Gly- (15)

B. Possible Steps in N-terminal Processing

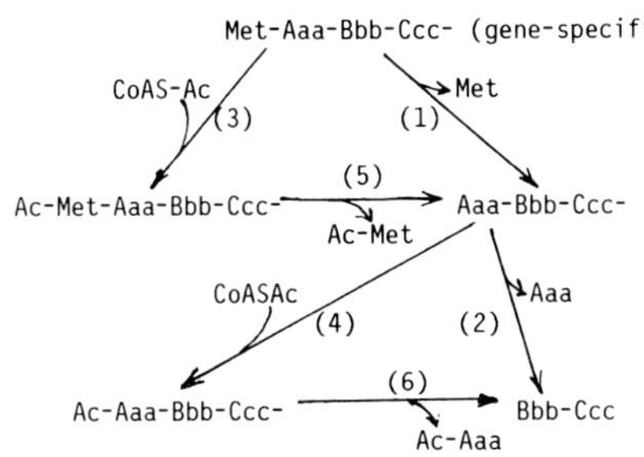

Fig. 1. Eukaryotic N-terminal processing of proteins without N-terminal signal peptide removal. A) Examples of some cases for which both cDNA and protein sequences are known. For aldolase, the two different protein sequences signify that the protein generally isolated from cells homogenized in dilute buffer has a free N-terminal Pro, while in one case, when it was isolated from cells homogenized in the presence of 1 mM tosylfluoride, it was found to have a blocked N-terminus [16]. B) Processing steps that may be required to account for all the known cases. Each peptide shown may either be an intermediate or an end product of processing.

The specificity of the purified enzymes for AcMet-peptides has been confirmed (see Note Added in Proof). The first order rate constant for the cleavage of 22 AcMet-terminated peptides varying in length from dipeptide to undecapeptide (most are tri and tetrapeptides) varies about 60-fold, but there is no obvious correlation with the peptide length or the nature of the amino acid in positions 2 and 3. The only exception to that conclusion may be that the two poorest substrates are a dipeptide and a heptapeptide with Glu in position 2; if these are excluded from the comparison, the rate constant variation for the remainder is within a factor of 8, with amino acids as different as Gly, Ala, Arg, Asp, Phe, and Met in position 2. The undecapeptide AcMet-Asp-Glu—[21] still appears to be the best substrate, along with the tripeptides AcMet-Leu-Gly and AcMet-Gly-Gly. Of 19 tri- to decapeptides starting with AcX-, where X is either Ala, Asp, Cys, Gly, Phe, Ile, Leu, Ser, or Val, none gave any evidence for cleavage (see Note Added in Proof). We have also acetylated two *E. coli* proteins containing N-terminal Met [cyanase (23) and dihydrofolate reductase (24)] and tested them as substrates, without finding any release of AcMet.

Given the above facts, is this enzyme involved to a significant extent in protein biosynthesis? There are several reasons why we feel that any specific AcMet-peptide-cleaving enzyme must be considered in the N-terminal processing of proteins. The most obvious one is the fact that such an activity is required for the observed processing of actin to take place, be it co- or post-translationally [13]. In addition, a number of isolated observations and generalizations may make a strong case for such an enzyme as a major activity in the removal of the initiator methionine. As illustrated in Figure 1, when muscle aldolase was isolated from tissue macerated in the presence of 1 mM tosylfluoride, the enzyme was found to have a blocked N-terminus, which could be removed by CNBr-treatment [10]. Thus the in vivo form of muscle aldolase appears to be AcMet-Pro-His, instead of the normally isolated Pro-His form of the enzyme, and if that is true for aldolase it could also apply to a number of other muscle proteins. Such an extrapolation might derive some indirect support for the observation that up to 80% of all soluble proteins are acetylated in ascites cells, for example [25].

It is also relevant to point out that the AcMet-peptide hydrolase was found to be 95% inhibited by 1 mM tosylfluoride; by that criterion it *could* be the enzyme responsible for the exposure of N-terminal Pro in the commonly isolated form of aldolase. Sherman and coworkers [26] have concluded that the initiator Met is removed from yeast cyt c isoforms and from proteins in other cells (by a putative Met-aminopeptidase) only if the second residue has a side chain with a small radius of gyration. To test this conclusion, they prepared mutants with an Ile residue in position 2, predicting that the Met would not be removed. That prediction was confirmed, but the isolated expressed product was not cyt c with a free Met-Ile N-terminus, but rather the acetylated form Ac-Met-Ile. Does this mean that the yeast aminopeptidase is incapable of removing the Met when Ile is in position 2, or does it mean that the Met-Ile- N-terminus is a unique substrate for a yeast transacetylase? Such a transfer of a set of specificity determinants from the aminopeptidase to the transacetylase actually has some interesting possibilities and leads to a model that is quite consistent with the available data. If we assume that the aminopeptidase is active regardless of the nature of residues 2 and 3, but that the transacetylase requires a bulky residue in positions 2 and/or 3, then some sequences would become resistant to the aminopeptidase because they are acetylated, while others would remain unacetylated and have their Met removed by this enzyme. With the bulky residues still available in the new position 2 and/or 3, the new N-terminus would be acetylated and, depending on the specificity of the acetylamino acylpeptide hydrolase(s), the final products could have either free or acetylated N-termini.

Such a model is consistent with the rules proposed by Sherman et al. [26] and also with the almost universal presence of bulky residues in positions 2 and/or 3 of acetylated proteins (Figs. 1 and 2) [16]. Clearly, the available data do not permit us to select a definite model at this time. Furthermore, it seems prudent at this stage not to assume that the signals and the machinery for N-terminal processing in lower (yeast) and higher eukaryotes (plants and animals) are the same. The cytochrome c case is a good illustration of that point; essentially all the animal cyts c contain an AcGly-, and higher plants have an AcAla- N-terminus, while the fungal cytochromes c have free, nonacetylated N-termini [27]. A similar case involving a cytosolic protein is illustrated in Figure 2 for enolases. All animal and plant enolases that have been isolated and characterized have been reported to have blocked N-termini, and for rabbit and salmon the blocked N-terminus has been identified as AcAla and AcSer, respectively [28]. Judging from the cDNA-derived sequences of human and rat enolases, the only N-terminal trimming that has taken place is removal of the Met, just as in the case of yeast enolase. But whereas residue 2 in the animal enolases (Ala or Ser) is acetylated, the yeast N-terminal Ala remains unblocked. With the extensive identity or similarity

							5					10					15
	Yeast		Ala	Val	Ser	Lys	Val	Tyr	Ala	Arg	Ser	Val	Tyr	Asp	Ser	Arg	Gly
(cDNA)	Yeast	Met	---	---	---	---	---	---	---	---	---	---	---	---	---	---	---
	Rabbit		Ac-Ala	Met	Gln	---	Ile	Phe	---	---	Glu	Ile	Leu	---	---	---	---
	Salmon		Ac-Ser	Ile	Thr	---	Ile	His	---	---	Glu	Ile	Leu	---	---	---	---
(cDNA)	Human	Met	Ser	Ile	Leu	---	Ile	His	---	---	Glu	Ile	Phe	---	---	---	---
(cDNA)	Rat	Met	---	Ile	Leu	---	Ile	His	---	---	Glu	Ile	Phe	---	---	---	---

						20					25					30	
	Yeast	Asn	Pro	Thr	Val	Glu	Val	Glu	Leu	Thr	Thr	Glu	Lys	Gly	Val	Phe	(12)
(cDNA)	Yeast	---	---	---	---	---	---	---	---	---	---	---	---	---	---	---	(11)
	Rabbit	---	---	---	---	---	---	Asp	---	His	---	Ala	---	---	Arg	---	(28)
	Salmon	---	---	---	---	---	---	Asp	---	Tyr	---	Ala	---	---	Arg	---	(28)
(cDNA)	Human	---	---	---	---	---	---	Asp	---	Phe	---	Ser	---	---	Leu	---	(29)
(cDNA)	Rat	---	---	---	---	---	---	Asp	---	Tyr	---	Ala	---	---	Leu	---	(30)

Fig. 2. N-terminal sequences of enolase from different sources. All sequences are compared with the yeast enolase sequence (top line); - - - in a given position means that that residue is identical to the corresponding one in yeast.

observed for the enolase sequences, the only reasonable deduction at this stage is that the acetylation signals and/or enzymes in yeast are different from those in animal cells.

Figure 2 [11,12,28–30] illustrates the rather frustrating aspect of the current state of available information relevant to this type of analysis of data bank information. For yeast, *both* cDNA and protein sequence data are available, permitting the firm conclusion that only Met has been removed from the encoded sequence to give the mature protein. For the mammalian enzymes, the sequences of two have been determined from the acetylated N-terminus by amino acid sequencing and of two others by cDNA sequencing. Thus we deduce from the four sequence alignments that the N-terminal processing steps for rabbit and salmon enolases are removal of Met followed by acetylation of residue 2, but we do not know this for a fact. It emphasizes again the need for having both protein and cDNA sequences before drawing conclusions about the processing; the frustrating part is that with the amazing wealth of sequence information obtained by either approach for a large number of proteins, the number of cases where both methods have been used for a single messenger/protein is still quite small. It goes without saying that the key question remains unanswered: did the established (yeast) or deduced (rabbit and salmon) removal of Met in the enolases in Figure 2 take place through the action of a Met aminopeptidase or through the action of a AcMet-protein-cleaving enzyme after acetylation of the Met in the precursor? As far as we can judge, this is probably an open general question at the moment. Based on the previous two discussions at this symposium, a picture emerges that suggests AcMet-removal is a post-translational event (for actin at least) and that the Met-aminopeptidase is responsible for the cotranslational N-terminal trimming. It would seem that a good deal of additional information is needed to establish this picture as a fact. The AcMet-removing enzyme discussed here certainly could be a candidate for the post-translational trimming; its specificity for AcMet- is an attractive feature, but at this stage it is not at all clear that it will act on N^{α}-acetylated proteins. The two protein substrates (*E. coli* proteins) that gave negative results may be totally wrong for all kinds of reasons, and proper substrates, such as, for example, the unprocessed actin precursor, need to be explored. The enzyme exists and must have a function.

There are actually other acetylamino acyl-peptide-cleaving enzymes that are candidates for the catalysis of AcMet-removal [18–20]. They are relatively nonspecific, and short peptides (3–4 amino acids) are better substrates than longer ones. The enzyme reported recently by Orfeo and Meyer [20], a cytosolic, high-MW (350,000) hydrolase found in a variety of vertebrate tissue, is particularly attractive. It has a micromolar Km and high V_{max} for peptides with AcMet- and AcCys- as N-terminals, millimolar Km and high V_{max} for Ac-Ala-peptides, millimolar Km and low V_{max} for peptides with AcGly, Gln, Leu, Phe, Tyr, and Val, and no activity for peptides with Ac-Asp, Glu, Arg, and Trp; it thus could, as the authors proposed, satisfy the requirements of the enzyme that carries out the trimming of both type I and type II actins [13].

A final point about the specific AcMet-peptide hydrolase discussed here. If this enzyme is a significant participant in eukaryotic protein synthesis, it should be present in other organisms as well. We have looked for it in bean sprouts and in yeast, using the standard HPLC-assay for AcMet-production [21], supplemented by a paper chromatography assay by which free amino acids released from the C-terminal end by carboxypeptidase can be detected. In yeast, the carboxypeptidase activity is so high that it has been impossible in preliminary experiments to establish unequivocally that AcMet is indeed produced from AcMet-peptides by a AcMet-peptide-cleaving enzyme rather than by carboxypeptidase. In the plant tissue, however, in spite of both contaminating carboxypeptidase and acylase (cleaving AcMet to free acetate and Met), it appears clear that the enzyme is present [Wen-Long Bai, personal communications].

The other purpose of this presentation is to consider the possible long-range specificity determinants involved in the decision of whether a protein is to have its initiator Met removed and then have the new N-terminus acetylated or not. The model we are considering proposes that a secondary structure, involving a rather long sequence (from 40–50 amino acids for cotranslational to the entire protein for post-translational modifications) may be involved, rather than only the first few N-terminal residues. The general basis for our model is that the first few residues may well determine some minimum specificity requirement, but that the display of that signal for "reading" and expression through the various processing steps requires specific secondary structure dispositions of the entire relevant part of the amino acid sequence (50 residues for cotranslational; all residues for post-translational modifications?). The main reason for making this extension of the specificity region is our recent report that a computer "learning" program distinguished between acetylated and nonacetylated sequences in the data bank and properly identified new sequences as falling into one group or the other by comparing the first 40 residues for their statistical assignment as secondary structure (helix, sheet, or turns) formers (+), breakers (−), or neutrals (0) [31]. The major distinguishing features between the two groups appear in the sequence ranges 1–10, 16–24, and 30–40. Visual inspection of groups of the 250 sequences used in the computer comparison does not reveal any obvious difference regarding specific amino acids or sequences, so the precise nature of the distinguishing properties has not yet been determined. At this stage it appears that the specificity determinants for N-terminal processing may be akin to the signal peptides, which will be discussed later in this symposium: they can clearly be defined according to certain patterns of charge and hydrophobic character, but no consensus sequence can be ascribed to them.

Another reason for suggesting a broader spectrum of specificity determinants is based on the fact that all our current attempts to derive simple rules seem to run into exceptions to the rules as more data become available. As already discussed above, there may be many reasons for these exceptions when we consider that for many cases we do not know where and when (co- or post-translationally) the N-terminal modifications took place or to what extent different organisms may use different signals for the processing. In addition, artifacts of isolation may give misleading results, as illustrated by the aldolase example discussed above.

The final point in this discussion is that in our attempts to establish rules about the specificity of N-terminal processing, we need to remind ourselves that, in spite of the impressive wealth of data now available through the data banks, we are probably looking at less than 10% of all proteins. Certainly, if we limit our law-making to well-documented cases in a single or in closely related organisms, the sample of known cases from which laws could be derived would be a rather meager one. It therefore seems prudent to keep the models as flexible as possible at this time and to consider the possibility that, as for so many co/post-translational modifications, secondary/tertiary/quarternary structure determines the reactivity of a certain site as much as does the immediate primary sequence around that site.

It seems fair to conclude at this time that both Met-aminopeptidases and AcMet-peptide hydrolases exist along with CoA-SAc-protein transacetylases, in at least some eukaryotic cells, and thus that the machinery for the reactions in Figure 1 is available. What the actual pathway is for the in vivo Met-removal and N-terminal acetylation for a given protein is essentially an unknown, and the specificity signals that determine whether or not the processing is to take place at all and by which pathway it is to proceed are equally enigmatic.

ACKNOWLEDGMENTS

We thank Theresa Domany for her patient and expert contributions to the preparation of this manuscript.

This work was supported in part by grants from NIH (GM-31305) and from the Robert A. Welch Foundation (AU-916).

REFERENCES

1. Capecchi MR: (1966) Proc Natl Acad Sci USA 55:1517–1524, 1966.
2. Adams JM, Capecchi MR: Proc Natl Acad Sci USA 55:147–154, 1966.
3. Waller JP: J Mol Biol 7:483–496, 1963.
4. Adams JM: J Mol Biol 33:571–589, 1968.
5. Ben-Bassat A, Baner K, Chang SY, Myambo K, Boosman A, Chang S: J Bacteriol 169:751–757, 1987.
6. Miller CG, Strauch KL, Kukral AM, Miller JL, Wingfield PT, Mazzei G, Werlan R, Graber P, Movva NR: Proc Natl Acad Sci USA 84:2718–2722, 1987.
7. Wold F: Annu Rev Biochem 50:783–814, 1981.
8. Putkey JA, Carroll SL, Means AR: Mol Cell Biol 7:1549–1553, 1987.
9. Tolan DR, Amsden AB, Putney SD, Urdea MS, Penhoet EE: J Biol Chem 259:1127–1131, 1984.
10. Lebherz HG, Bates OJ, Bradshaw RA: J Biol Chem 259:1132–1135, 1984.
11. Holland MJ, Holland JB, Thill GP, Jackson KA: J Biol Chem 256:1385–1395, 1981.
12. Chin CCQ, Brewer JM, Wold F: J Biol Chem 256:1377–1384, 1981.
13. Rubenstein P: This volume.

14. McReynolds L, O'Malley BW, Nisbet AD, Fothergill JE, Givol D, Fields S, Robertson M, Brownlee GG: Nature 273:723–728, 1978.
15. Narita K, Ishii J: J Biochem 52:367–373, 1962.
16. Driessen HPC, deJong WW, Tesser GI, Bloemendal H: CRC Crit Rev Biochem 18:281–325, 1985.
17. Bradshaw RA: This volume.
18. Gade W, Brown JL: J Biol Chem 253:5012–5018, 1978.
19. Jones WM, Manning JM: Biochem Biophys Res Commun 126:933–940, 1985.
20. Orfeo T, Meyer WL: Fed Proc 45:1712, 1986.
21. Radhakrishna G, Wold F: J Biol Chem 261:9572–9575, 1986.
22. Wool IG: Annu Rev Biochem 48:719–754, 1979.
23. Chin CCQ, Anderson PM, Wold F: J Biol Chem 258:276–282, 1983.
24. Stone D, Phillips AW, Burchall JJ: Eur J Biochem 72:613–624, 1977.
25. Brown JL, Roberts WK: J Biol Chem 251:1009–1014, 1976.
26. Sherman F, Stewart JW, Tsunasawa S: Bio Essays 3:27–31, 1985.
27. Dickerson RE, Timkovich R: In: Boyer PD (ed): "The Enzymes," Vol XI. New York: Academic Press Inc., 1975, pp 397–547.
28. Chin CCQ, Wold F: Biosci Rep 5:847–854, 1985.
29. Giallongo A, Feo S, Moore R, Croce CM, Showe LC: Proc Natl Acad Sci USA 83:6741–6745, 1986.
30. Sakimura K, Kushiya E, Obinata M, Takahashi Y: Nucleic Acids Res 12:4365–4378, 1985.
31. Augen J, Wold F: Trends Biochem Sci 11:494–497, 1986.

NOTE ADDED IN PROOF

Recent studies on the pure AcMet-peptide hydrolase (7000-fold purification) using improved HPLC-assays with [^{14}C]-Ac-peptide substrates and [^{3}H]-Ac-amino acids as product reference compounds show that some of our conclusions reported here and in a previous publication [21] are wrong. The native enzyme appears to have an M_r of 240,000 and be composed of identical 80,000 dalton subunits. More importantly, the enzyme is *not* specific for AcMet-peptides; other Ac-amino acids (e.g., AcAla and AcGly) are also cleaved from acetylated peptides. The specificity appears to be complex, and more data will be needed to fully elucidate it, but clearly this is not as we have claimed, a specific AcMet-peptide hydrolase. Based on both the M_r and our limited specificity data, our enzyme appears to be similar to the activity Orfeo and Meyer [20] have described in muscle as well as in other tissues. Even without any evidence for a specific AcMet-peptide hydrolase, we feel that the questions raised in this report about the possible role of such an enzyme are valid and should be included in the overall consideration of N-terminal protein processing.

Molecular Biology of Intracellular Protein Sorting and Organelle Assembly 117–121 (1988)

Discussion Summary: Lysosomal Targeting

Suzanne R. Pfeffer, Catherine Nolan, and Tom H. Stevens

Department of Biochemistry, Stanford University School of Medicine, Stanford, California 94305 (S.R.P.); Department of Biochemistry, St. Louis University School of Medicine, St. Louis, Missouri 63104 (C.N.); and Institute of Molecular Biology, University of Oregon, Eugene, Oregon 97403 (T.H.S.)

Proteins bound for the cell surface, lysosomes, and secretory storage granules (in cells that store their secretory products) share a common pathway of intracellular transport. Thus, after translocation across the membrane of the endoplasmic reticulum (ER), this mixture of proteins is transported together from the ER to the *cis*, medial, and then *trans* Golgi, and it is only after reaching the *trans* Golgi that their paths diverge.

Work from a number of laboratories suggests that transport of proteins from the Golgi complex to the cell surface is a nonselective, bulk flow process, whereas the diversion of proteins to lysosomes or secretory storage granules is signal-mediated [see 1 for review]. The signals used to target a number of proteins to lysosomes are now fairly well characterized in animal cells and have recently been identified in yeast (see below). It is likely that the signals that target proteins to secretory granules will be elucidated in the next few years. Despite the identification of lysosome-specific signals and, in animal cells, of receptors that capture modified lysosomal enzymes for transport, virtually nothing is known about how such receptors are subsequently collected into appropriate transport vesicles, how these transport vesicles bud, and how they then recognize their select targets. A complete understanding of the lysosomal sorting problem will require biochemical and genetic investigations aimed at unravelling these fundamental questions. This workshop summary will be divided into two parts: first, an overview of lysosomal targeting in animal cells, followed by a summary of our understanding of vacuolar targeting in the smaller eukaryote, *Saccharomyces cerevisiae*.

In animal cells, the targeting of proteins to lysosomes has been studied in great detail [see 2–5 for review]. Upon arrival in the Golgi apparatus, newly synthesized lysosomal enzymes are recognized by an N-acetylglucosaminyl phosphotransferase that adds GlcNAc-P to terminal alpha-1,2-linked mannose residues. A specific phosphodiesterase then removes the GlcNAc, to uncover the man-6-P residues that enable lysosomal enzymes to bind to man-6-P receptors in the Golgi apparatus. Receptor-bound hydrolases are then transported to an acidic prelysosomal compartment, within which the ligand is

Received March 30, 1987.

released for transport to lysosomes, and the free receptor recycles back to the Golgi apparatus.

Two types of man-6-P receptors have been identified. One has been purified from a number of sources and is a single polypeptide of 215,000 daltons. However, the 215-kD receptor is not present in a number of cell lines that adequately target their hydrolases to lysosomes [6]. Such cells were recently found to contain a second man-6-P receptor glycoprotein, comprised of three 46-kD subunits [7]. This receptor, which appears to be present in many cell types, requires divalent cations for ligand binding, and has thus been termed man-6-P receptor "CD" (cation-dependent) to distinguish it from the 215-kD cation-independent man-6-P receptor "CI" [7]. The two receptors display slight differences in their pH dependence for ligand binding [8]; however, nothing is yet known about why cells contain both receptors and whether or not these proteins have distinct roles within the cell.

While the identification of such sorting signals and cognate receptors is gratifying, it appears that certain proteins are targeted to lysosomes by an entirely different mechanism. Ira Mellman described his laboratory's work on lysosomal membrane glycoproteins that never acquire man-6-P residues [9]. This family of glycoproteins is highly sialylated. Mellman et al. used several mutant cell lines lacking specific glycosyltransferases to demonstrate that terminal glycosylation has no discernible effect on the stability of these proteins or apparent lysosome function. As has been appreciated for several years, hepatocytes, Kupffer cells, and leukocytes from patients with I-cell disease lack the ability to add man-6-P residues, yet contain normal levels of intracellular lysosomal enzymes. Clearly, another targeting mechanism must exist to direct these proteins to lysosomes. The solution to this puzzle awaits further investigation.

INTRACELLULAR TRANSPORT OF MAN-6-P RECEPTORS

At steady state, the 215-kD man-6-P receptor is found in the Golgi apparatus and in endosomes; a small fraction is present on the cell surface. Recent experiments suggest that the 215-kD receptor is in rapid equilibrium between these intracellular locations, since at least 90% of the receptor contacts extracellularly administered anti-receptor antibodies within 90 minutes [10,11]. However, several reports have indicated that the receptor may accumulate in the *cis* Golgi under conditions in which synthesis of lysosomal hydrolases or the man-6-P recognition marker is blocked [12,13]. These findings could only be explained if the 215-kD man-6-P receptor was selectively retained in the Golgi complex and only exported after ligand binding.

In order to study the intracellular transport of the man-6-P receptor, Suzanne Pfeffer employed an endosome-specific radiolabeling procedure to biochemically detect the man-6-P receptor when present in endosomes. If the 215-kD man-6-P receptor accumulates in the Golgi apparatus in the absence of lysosomal hydrolase synthesis, it should be correspondingly depleted from endosomes after a period of cycloheximide treatment. To test this, cells were treated with cycloheximide for up to 6 hours to permit complete transit of newly synthesized lysosomal hydrolases to lysosomes. The transferrin receptor was used as an internal control; the endosomal level of the transferrin receptor would not be expected to change in the presence of cycloheximide, since this protein has been shown to recycle at the same rate in this cell line in the absence of protein synthesis [14]. The experiments revealed that the endosomal level of both receptors was unchanged up to 6 hours after addition of the protein synthesis inhibitor. Additional control experiments demonstrated that the equilibrium of the receptor between cellular compartments, as measured by accessibility of the man-6-P receptor to extracellularly administered receptor antibodies, was unchanged after cycloheximide treatment. Taken together, these findings strongly suggest that the man-6-P receptor cycles continuously between intracellular compartments, with or without bound ligand.

Mellman has shown that antibody-crosslinking diverts Fc receptors to lysosomes, where they are degraded [15]. Catherine Nolan and William Sly described antibody-induced impairment of the function of the 215-kD man-6-P receptor, with inhibition of both transport of newly synthesized acid hydrolases and of endocytosis of exogenous ligands. Antibodies to the receptor inhibited its function, not by interfering with ligand binding, but by inhibiting its normal recycling, probably by receptor crosslinking. The specificity of such antibodies makes these useful reagents with which to further study the transport of acid hydrolases.

Analysis of the receptors and signals used to target lysosomal enzymes is clearly important, in part because they are one step closer to the sorting machinery. In order to study the molecular basis of the vesicular transport of the man-6-P receptor between intracellular compartments, it will be important to reconstitute this process in a cell-free system. With this goal in mind, a complementation scheme was devised by one of us (S.P.) to measure the arrival of the man-6-P receptor from (potentially) the Golgi complex to a prelysosomal organelle. The prelysosomal compartment of cells lacking the 215-kD man-6-P receptor was filled with lactoperoxidase (LPO), and a crude extract was prepared. This extract was then mixed with an extract from cells that contain the man-6-P receptor and incubated at 37°C in the presence of ATP. At the end of the incubation, mixtures were cooled to 4°C, and ^{125}I and H_2O_2 were added to initiate an LPO-mediated iodination reaction within the prelysosomal organelles. The man-6-P receptor should have become iodinated if transport between organelles had occurred, as was observed after specific receptor immunoprecipitation. Numerous additional control experiments will be necessary to determine from which compartment the receptor originated. However, these preliminary experiments suggest that it will indeed be possible to reconstitute the transfer of the man-6-P receptor between intracellular compartments.

VACUOLAR TARGETING

The second half of the workshop was devoted to recent advances in our understanding of how proteins are sorted to the yeast lysosome-like vacuole [16]. Because of the readily manipulated genetic and transformation systems available in yeast, this simple eukaryote is ideal for addressing certain questions in protein sorting. The yeast vacuole is a low-pH compartment containing a large number of hydrolytic glycoproteins and is thus analogous to the mammalian lysosome [17]. Vacuolar glycoproteins transit the ER and Golgi apparatus en route to the vacuole [18]. However, one major difference between yeast and mammalian cell protein sorting is that whereas soluble mammalian lysosomal enzymes require Asn-linked carbohydrate for sorting, yeast-soluble vacuolar proteins are sorted and transported to the vacuole with high efficiency, even when Asn-linked glycosylation is blocked with tunicamycin [18]. Therefore, one of the primary goals for laboratories studying yeast vacuolar protein sorting has been to elucidate the nature of this targeting signal(s). A second important goal has been the elucidation of the cellular machinery in yeast involved in recognizing these targeting signals. Finally, the issue of the sorting and transport of yeast vacuolar membrane proteins was raised and discussed.

Luis Valls et al. [19] described a genetic analysis of the vacuolar targeting signal for the yeast vacuolar glycoprotein carboxypeptidase Y (CPY). Mutations in the CPY structural gene *PRC1*, which cause the CPY precursor (proCPY) to be missorted and secreted, all map to the extreme amino terminus (aa21-30) of proCPY, adjacent to the CPY signal sequence (aa1-20). The efficiency of sorting of these mutant precursors to the vacuole varies from 50% to as little as 5% of the newly synthesized CPY. The only point mutation isolated was a change of Gln→Lys at residue 24, and about 90% of this mutant proCPY was missorted to the cell surface. Interestingly, it was found that all the mutant, secreted CPY precursors could be activated by trypsin to the enzymatically active CPY species with an efficiency equal to that found for activation of the wild-type CPY precursor. Thus these mutations have altered the vacuolar targeting signal in CPY without altering the overall conformation of the enzyme.

Vytas Bankaitis et al. [20] took a gene fusion approach to the question of the vacuolar targeting signals in proCPY. Chimeric genes were constructed with the cytoplasmic form of the yeast protein invertase (inv) fused to the carboxy terminus of CPY. Fusions containing the first 20 amino acids of CPY (CPY_{20}-inv) fused to inv result in the efficient secretion of the fusion protein to the cell surface. In larger fusions, CPY_{30}-inv, 65% of the fusion protein was found in the vacuole. Interestingly, with as few as 50 amino acids from CPY, CPY_{50}-inv, the fusion protein was found exclusively in the vacuole. In addition, deletion of residues 21–50 from the *PRC1* gene resulted in the secretion of the encoded mutant proCPY. Taken together with the results of Valls et al. [19], it is reasonable to conclude that the protein domain of amino acid residues 21–30 (or 21–50) is necessary and sufficient to serve as a vacuolar targeting signal. Therefore it is apparent that CPY carries two adjacent topogenic sequences, the signal sequence at the extreme amino terminus, followed by the vacuolar sorting determinant in the adjacent region of the protein.

Joel Rothman et al. [21] and Vytas Bankaitis et al. [22] described two different genetic selections that have identified mutant cells that fail to sort newly synthesized vacuolar proteins. These vacuolar protein mislocalization mutations, *VPL* [21] and *VPT* [22], cause cells to secrete a large percentage of proCPY [21,22] and at least one other soluble vacuolar protein [21]. However, the number of identified complementation groups has greatly expanded, since there are now 35 *VPT* and 16 *VPL* complementation groups, and significant overlap between these two classes of mutants is anticipated. A preliminary analysis indicates that vacuolar membrane proteins are transported normally in *VPL* mutant cells [21], suggesting that there may be another as yet unidentified class of proteins involved in the sorting of vacuolar membrane proteins. It was clear from the presentations and discussions that it will be important in the very near future to determine which of the *VPL* and *VPT* gene products are intimately involved in the sorting reaction (e.g., identification of the CPY sorting receptor and then elucidation of its function in vitro).

Joel Rothman et al. also discussed their progress toward elucidating the sorting mechanism of vacuolar membrane proteins. The vacuolar integral membrane protein, dipeptidyl aminopeptidase B (DPAP-B), has been found to undergo Asn-linked glycosylation and requires the secretory pathway for its transport to the vacuole. A genetic scheme has been developed that will allow the isolation of *cis*- and *trans*-acting mutations that cause cells to missort DPAP-B into the late secretory pathway [16]. Mutagenesis of the DPAP-B structural gene will allow a determination of which portions of the protein (e.g., cytoplasmic tail or transmembrane domain) are required for its localization to the vacuole. Mutations unlinked to the DPAP-B structural gene will probably define genes whose products are involved in vacuolar membrane protein sorting, similar to the genes (*VPL* and *VPT*) that are required for soluble vacuolar protein sorting.

The material presented in this workshop demonstrated that in vitro assays as well as genetic approaches constitute powerful and complementary methods for elucidating the molecular mechanisms of intracellular protein sorting.

REFERENCES

1. Pfeffer SR, Rothman, JR: Annu Rev Biochem 56:829, 1987.
2. Kornfeld S: J Clin Invest 77:1, 1986.
3. von Figura K, Hasilik A: Annu Rev Biochem 55:167, 1986.
4. Sahagian GG: In Parent B, Olden K (eds): "Recent Research on Vertebrate Lectins." Adv Cell Biol Monographs, Van Nostrand Reinhold Company Inc., New York, pp 46–64, 1987.
5. Sly WS, Fischer HD: J Cell Biochem 18:67, 1982.
6. Gabel CA, Goldberg DE, Kornfeld S: Proc Natl Acad Sci USA 80:775, 1983.
7. Hoflack B, Kornfeld S: Proc Natl Acad Sci USA 82:4428, 1985.
8. Hoflack B, Kornfeld S: J Biol Chem 262:123, 1987.
9. Lewis V, Green SA, Marsh M, Vihko P, Helenius A, Mellman I: J Cell Biol 100:1839, 1985.
10. Sahagian G: Biol Cell 51:207, 1984.
11. von Figura K, Gieselmann V, Hasilik A: EMBO J 3:1281, 1984.
12. Brown WJ, Farquhar MG: Proc Natl Acad Sci USA 81:5135, 1984.
13. Brown WJ, Constantinescu E, Farquhar MG: J Cell Biol 99:320, 1984.
14. Ciechanover A, Schwartz AL, Dautry-Varsat A, Lodish HF: J Biol Chem 258:9861, 1983.
15. Mellman I, Plutner H: J Cell Biol 98:1170, 1984.
16. Rothman JH, Stevens TH: In Das RC, Robbins OW (eds): "Protein Transfer and Organelle Biogenesis." New York: Academic Press, in press, 1988.
17. Wiemken A, Schellenberg M, Urech K: Arch Microbiol 123:23, 1979.
18. Stevens TH, Esmon B, Schekman R: Cell 30:439, 1982.
19. Valls LA, Hunter CP, Rothman JH, Stevens TH: Cell 48:887, 1987.
20. Johnson LM, Bankaitis VA, Emr SD: Cell 48:875, 1987.
21. Rothman JH, Stevens TH: Cell 47:1041, 1986.
22. Bankaitis VA, Johnson LM, Emr SD: Proc Natl Acad Sci USA 83:9075, 1986.

Molecular Biology of Intracellular Protein
Sorting and Organelle Assembly 123–126 (1988)

Discussion Summary: Protein Processing Enzymes

Reid Gilmore and Mark O. Lively

Department of Biochemistry, University of Massachusetts Medical Center, Worcester, Massachusetts 01605 (R.G.); and Department of Biochemistry, Bowman Gray School of Medicine, Wake Forest University, Winston-Salem, North Carolina 27103 (M.O.L.)

Cotranslational and post-translational modification, or processing, of polypeptides is essential for the maturation of a substantial proportion of proteins within a cell before they become functional molecules at their proper sites of action. Because the number of protein processing events known is quite large [1], the workshop concentrated on two protein processing reactions that have received considerable attention during the past several years: 1) the removal of amino terminal signal (leader) sequences by the enzymes known as signal peptidases; and 2) the acylation of proteins with fatty acids.

EUKARYOTIC SIGNAL PEPTIDASE

The workshop began with a brief presentation by Emily Evans (Rockefeller University) concerning the isolation of signal peptidase from canine pancreas. The purification procedure described by Evans and coworkers [2] resulted in the isolation of a protein complex consisting of six polypeptides with apparent molecular masses of 25, 23, 22, 21, 18, and 12 kD. The 22- and 23-kD proteins were shown to be glycosylated based upon their ability to bind concanavalin A. Digestion of the signal peptidase complex with endoglycosidase H eliminated the 22- and 23-kD proteins and resulted in the appearance of a single 19-kD digestion product. The comigration of the 22- and 23-kD subunits following deglycosylation raised the question of whether these two subunits are distinct polypeptides or instead differ only in the structure of their asparagine-linked oligosaccharides. Evans reported that the 22- and 23-kD polypeptides have identical N-terminal amino acid sequences, so it seems certain that these two glycoproteins differ only in the degree of oligosaccharide processing. Further support for this conclusion was provided by Cleveland peptide mapping analysis [3] using chymotrypsin, which suggested that the 22- and 23-kD subunits are identical (Grosse and Gilmore, unpublished data). Thus the signal peptidase complex from canine pancreas contains five protein subunits in apparently equal quantities, one of which is heterogenously glycosylated. The multiple glycosylated polypeptides resolved by SDS polyacrylamide gel electro-

Received March 31, 1987.

phoresis may indicate a variability in the extent of mannose trimming of the asparaginyl core oligosaccharide as observed for ribophorin I [4].

Several lines of evidence support the conclusion that the five subunits of purified canine pancreas signal peptidase constitute a physiologically relevant complex. They copurified with signal peptidase activity during numerous purification steps, including ion exchange chromatography, gel filtration, and sucrose density gradient centrifugation. A purification of the signal peptidase complex can also be achieved by lectin affinity chromatography on concanavalin A Sepharose. The same polypeptide profile is obtained by eluting this column with alpha-methyl mannoside, although only two of the six proteins are glycosylated. This observation provides additional support for the existence of a signal peptidase complex: the nonglycosylated subunits must bind to the column via interaction with a glycoprotein, since nonglycosylated proteins do not bind concanavalin A directly. Previously, Evans and coworkers [2] had proposed that a single subunit of the complex may be responsible for the signal peptidase activity. Evidence for this proposal has not yet been obtained, so additional research will be required to identify the subunit(s) that contains the proteolytic active site and to determine what functions, if any, are associated with the other subunits.

Mark Lively described the isolation of signal peptidase from hen oviduct microsomal membranes. The hen oviduct enzyme was purified to near homogeneity after detergent solubilization of oviduct microsomal membranes that had been stripped of peripheral membrane proteins by treatment with ice-cold sodium carbonate at pH 11.5 [5,6]. Purified hen oviduct signal peptidase consists of only three proteins with apparent molecular weights of 23, 22, and 19 kD. These three proteins are the only proteins that are bound specifically to lentil lectin (*lens culinaris* agglutinin) Sepharose (Baker and Lively, unpublished data). The 23- and 22-kD proteins were shown to be glycosylated, since they bind concanavalin A following electrophoretic transfer to nitrocellulose. Digestion of the enzyme with endoglycosidase H or N-glycanase, a glycosidase specific for all classes of asparaginyl oligosaccharides, converted the 23- and 22-kD glycoproteins to a single 19-kD protein band on SDS acrylamide gel electrophoresis.

Hence, hen oviduct signal peptidase, unlike the canine pancreas enzyme, is not purified as a complex of six proteins. However, the canine and hen enzymes each have in common a pair of glycoproteins that appear to be differentially glycosylated versions of a single polypeptide. Perhaps polypeptides corresponding to the additional subunits of the canine pancreas complex are dissociated from the hen enzyme by the carbonate treatment procedure used during its purification. Additional research is required to determine whether other proteins may be associated in a complex with hen oviduct signal peptidase in vivo.

Although several laboratories have attempted to identify inhibitors of the eukaryotic signal peptidase, specific active site inhibitors have not yet been identified. Joyce Sutcliffe (Pfizer Central Research, Groton, CT) discussed the importance of identifying inhibitors of prokaryotic leader peptidase that do not inhibit the eukaryotic signal peptidase, since such inhibitors could potentially be clinically useful antibiotics. She reported that synthetic peptide analogs of the leader sequence of M13 procoat longer than 16 amino acid residues appear to inhibit leader peptidase.

CLEAVED AND UNCLEAVED SIGNAL SEQUENCES

Evans described a recent observation concerning the processing of yeast prepro-alpha-factor, a 165-residue polypeptide precursor of the alpha-mating pheromone. Prepro-alpha-factor contains a 20-residue hydrophobic sequence at the amino terminus, which is the presumed signal sequence. Previous research by Julius and coworkers [7] has shown that the signal sequence of prepro-alpha-factor is not cleaved in vivo in yeast or in an in vitro translation system supplemented with canine pancreas microsomal membranes. Surprisingly, incubation of prepro-alpha-factor with the purified canine pancreas signal peptidase led to the appearance of an additional alpha-factor-specific polypeptide with a mobility slightly slower than that of prepro-alpha-factor [Evans, Waters, and Blobel, unpublished data]. A polypeptide with an identical mobility has been detected in yeast in vitro translocation systems [8,9]. Amino terminal sequencing of the novel alpha-factor product demonstrated that signal peptide cleavage had occurred between residues 19 and 20. An examination of this sequence using the criteria of Von Heijne [10] predicts a signal peptidase cleavage site at this position. The anomalous mobility of unglycosylated pro-alpha-factor relative to prepro-alpha-factor on SDS gels is not understood. Further research will be required to determine whether cleavage of the signal peptide occurs at this site in a homologous yeast translocation system in vitro.

The transforming growth factor-alpha (TGF-alpha) is a mitogenically active polypeptide hormone produced by acutely proliferating neoplastic cells. Analysis of the rat TGF-alpha cDNA [11] revealed that the TGF-alpha gene encodes a 159 amino acid residue prohormone (proTGF-alpha) that contains a 24-residue hydrophobic sequence at the amino terminus, a potential glycosylation site at asparagine 25, and a potential transmembrane spanning sequence located between amino acid residues 97 and 120. Translation of rat proTGF-alpha mRNA in vitro in the presence of microsomal membranes demonstrated that proTGF-alpha is synthesized as an integral membrane glycoprotein [12]. Unlike many membrane proteins with amino terminal signal sequences, the signal sequence of proTGF-alpha appears not to be removed upon membrane integration. Examination of the signal sequence revealed a single potential signal peptidase cleavage site between residues 22 and 23 based on the presence of small, neutral amino acids at positions -1 and -3 before the predicted site of cleavage [10]. The failure of signal peptidase to cleave at this site is presumably due to the abnormally short distance (three amino acids) between the hydrophobic core of the signal sequence and the potential Ala-Leu cleavage site [13].

FATTY ACID ACYLATION OF RAS PROTEINS

The attention of the workshop was next focused on the acylation of proteins with fatty acids. It is now clear that targeting of certain membrane proteins to their proper sites of action requires protein processing events that occur within the cytoplasm. Covalent addition of fatty acids to some proteins is responsible, at least in part, for their membrane localization. *RAS* genes encode proteins that play an essential role in growth regulation in cells, and it has been shown that specific acylation of these proteins with palmitic acid results in their localization on the inner surface of the plasma membrane [14]. This processing event is critical for these proteins, because membrane localization is required for expression of biological activity [15,16]. The fatty acid attachment sites of *RAS* proteins share a common amino acid sequence: Cys-Aaa-Aaa-Xxx, where Aaa represents an aliphatic amino acid, and Xxx is the C-terminal residue of the protein [17]. Palmitic acid is attached as a thioester to the cysteine residue [16,18]. While it is believed that palmitylation is at least in part responsible for membrane binding, the mechanism by which this modification occurs and results in specific localization of the acylated protein on the cytoplasmic face of the plasma membrane is not known.

Fuyuhiko Tamanoi (University of Chicago) discussed how studies of *RAS* genes in yeast may lead to an understanding of the function of fatty acylation of some proteins. He has isolated mutants of *Saccharomyces cerevisiae* defective in the processing of *RAS* proteins [19]. These mutants are known as *DPR1* and are characterized by the following: 1) accumulation of *RAS2* protein precursors in the cytoplasm; and 2) lower levels of *RAS2* proteins in the plasma membrane. However, the acylation reaction appears to take place in these cells, albeit at a reduced level. These results suggest that the major effect of the *DPR1* mutation is in a processing step that takes place prior to acylation. The exact nature of this processing event has not been defined. Although the nonacylated intermediate has an increased mobility upon SDS polyacrylamide gel electrophoresis, results of microsequence analysis of the precursor and its processing intermediate indicate that proteolysis of the amino terminus of the molecule does not occur [19].

Similar yeast mutants were described by Susan Michaelis (University of California at San Francisco) in her plenary lecture at the UCLA meeting [20]. These mutants, termed *RAM* [21], are believed to encode an enzyme responsible for modification and membrane localization of proteins containing the C-terminal Cys-Aaa-Aaa-Xxx sequence. The *RAM* and *DPR1* mutants have been shown by Michaelis to be the same [20]. The *RAM* protein may encode the fatty acid acyltransferase or possibly an enzyme required to modify *RAS* and other proteins including the yeast *a* factor pheromone precursor, so that they may be subsequently acylated. Further studies of these mutants will lead to a better understanding of the protein processing reactions that result in membrane localization as a result of acylation with fatty acids.

REFERENCES

1. Uy R, Wold F: Science 198:890, 1977.
2. Evans EA, Gilmore R, Blobel G: Proc Natl Acad Sci USA 83:581, 1986.
3. Cleveland DW, Fischer SG, Kirschner MW, Laemmli UK: J Biol Chem 252:1102, 1977.
4. Rosenfeld MG, Marcantonio EE, Hakimi J, Ort VM, Atkinson PH, Sabatini D, Kreibich G: J Cell Biol 99:1076, 1984.
5. Lively MO, Walsh KA: J Biol Chem 258:9488, 1983.
6. Baker RK, Bentivoglio GP, Lively MO: J Cell Biochem 32:75, 1986.
7. Julius D, Schekman R, Thorner J: Cell 36:309, 1984.
8. Waters MG, Blobel G: J Cell Biol 102:1543, 1986.
9. Hansen W, Garcia PD, Walter P: Cell 45:397, 1986.
10. Von Heijne G: Eur J Biochem 133:17, 1983.
11. Lee DC, Rose TM, Webb NR, Todaro GJ: Nature 313:489, 1985.
12. Teixido J, Gilmore R, Lee DC, Massague J: Nature 326:883, 1987.
13. Von Heijne G: J Mol Biol 173:243, 1984.
14. Sefton BM, Trowbridge IS, Cooper JA, Scolnick EM: Cell 31:465, 1982.
15. Weeks MO, Hager GL, Lowe R, Scolnick EM: J Virol 54:586, 1985.
16. Willumsen BM, Christensen A, Hubbert NL, Papageorge AG, Lowry DR: Nature 310:583, 1984.
17. Taparowsky E, Shimizu K, Goldfarb M, Wigler M: Cell 34:581, 1983.
18. Buss JE, Sefton BM: Mol Cell Biol 6:116, 1986.
19. Fujiyama A, Matsumoto K, Tamanoi F: EMBO J 6:226, 1987.
20. Michaelis S, Herskowitz I, Powers S, Wigler M: (in this volume).
21. Powers S, Michaelis S, Broek D, Anna-A. SS, Field J, Herskowitz I, Wigler M: Cell 47:413, 1986.

Molecular Biology of Intracellular Protein Sorting and Organelle Assembly 127–132 (1988)

Resolution of Distinct Steps in Mitochondrial Protein Import

Nikolaus Pfanner, Franz-Ulrich Hartl, and Walter Neupert

Institut für Physiologische Chemie, Universität München, D-8000 München 2, Federal Republic of Germany

The transport of proteins from the cytosol into mitochondria involves several distinct steps. Precursor proteins are synthesized on cytosolic polysomes; most carry aminoterminal presequences which contain sufficient information for the targeting of proteins to mitochondria. Hydrolysis of nucleoside triphosphates is necessary to keep the precursors in an import-competent conformation. The precursors bind to specific receptors on the mitochondrial surface and are then inserted into the outer membrane. Transport of precursors into or across the inner membrane takes place at translocation contact sites and requires the mitochondrial membrane potential. The presequences are cleaved off by the processing peptidase in the mitochondrial matrix. With several precursors, a second proteolytic processing step is also performed. Finally, proteins are assembled into multi-subunit complexes.

Key words: mitochondria, precursor proteins, receptors, contact sites

Two protein synthesizing systems are required for the biogenesis of mitochondria [for review, see 1–3]. A small number of proteins are coded for by the mitochondrial genome and are translated within mitochondria. The majority of mitochondrial proteins, however, are coded by nuclear genes and synthesized on cytosolic polysomes. The transport of proteins from the cytosol into mitochondria involves four essential features: 1) specific recognition of precursors by mitochondria; 2) translocation of proteins across the mitochondrial membranes; 3) sorting of proteins to one of the four mitochondrial compartments (outer membrane, intermembrane space, inner membrane, or matrix); and 4) functional assembly.

Several steps in the complex process of mitochondrial protein import have been resolved in the past 12 years. Our present knowledge is summarized in Table I. The details will be discussed in the following sections.

CYTOSOLIC PRECURSORS FOR MITOCHONDRIAL PROTEINS

In vivo and in vitro studies have shown that mitochondrial proteins, which are synthesized on cytosolic polysomes, can be released as water-soluble precursors into the

Received April 9, 1987.

TABLE I. Proposed Steps for the Import of Mitochondrial Precursor Proteins

1. Synthesis on free cytosolic polysomes.
2. Nucleoside triphosphate-dependent folding of precursors into an import-competent conformation.
3. Amino terminal presequences (or positively charged domains in precursors without presequences) serve as targeting signals.
4. Binding to receptor proteins on the mitochondrial surface.
5. Insertion into the outer membrane.
6. Import via translocation contact sites.
7. Membrane potential-dependent insertion into the inner membrane.
8. Proteolytic processing by the processing peptidase located in the matrix; in some cases a second proteolytic processing step is performed (either by the processing peptidase or by so far uncharacterized processing activities). Attachment of prosthetic groups.
9. Intermembrane space and inner membrane proteins: either complete import into the matrix and sorting from the matrix space, or lateral diffusion from contact sites to their functional destination.
10. Assembly into multisubunit complexes.
11. Exceptions: import independent of contact sites and membrane potential.
    a) Outer membrane proteins (which have no cleavable presequence) are directly inserted into the outer membrane. b) Apocytochrome ***c*** (no cleavable presequence) is translocated across the outer membrane; holocytochrome ***c*** is formed following covalent heme attachment.

cytosol. The precursors are then post-translationally imported into mitochondria [4–6]. Their transport in the cytosol occurs mainly in the form of higher molecular weight aggregates [7]. Many precursors are synthesized with amino terminal presequences that are about 20 to 70 amino acid residues in length and that are positively charged [1,2]. The presequences seem to carry sufficient information for the targeting of proteins to mitochondria [8–10]. This was first shown by studies with fusion proteins that consisted of a mitochondrial presequence joined to a nonmitochondrial protein.

PROTEINACEOUS RECEPTORS ON THE MITOCHONDRIAL SURFACE

The first step in the interaction of precursors with mitochondria involves binding to specific receptor proteins on the mitochondrial surface. This is shown by the following lines of evidence.

1. Pretreatment of mitochondria with low concentrations of proteases inhibits the import of precursors [11–14]. This treatment with proteases does not inhibit the mitochondrial membrane potential, or the transport of precursors from the outer into the inner membrane, or the intramitochondrial sorting of precursors [15,16].

2. The number of binding sites and their affinity constants have been determined for the binding of three different precursors to mitochondria: porin of the outer membrane [14,17]; cytochrome *c* of the intermembrane space [18]; and the ADP/ATP carrier of the inner membrane [19].

3. Precursors can be bound to mitochondria in the absence of a membrane potential ($\Delta\Psi$). After re-establishing $\Delta\Psi$, they are imported from the bound state [20,21].

Apocytochrome *c*, the precursor to (holo)cytochrome *c*, does not compete for the binding or import of any other protein tested so far [22,23]. The precursor to porin, however, seems to compete for the specific binding of ADP/ATP carrier to mitochondria [17]. In summary, mitochondria apparently carry more than one class of import receptors.

On the other hand, even precursors destined for distinct submitochondrial locations may use the same receptor sites.

TRANSLOCATION CONTACT SITES

Sites of close contact between the outer and inner mitochondrial membranes were described in electron microscopic studies 20 years ago [24]. Recently, it could be shown that import of mitochondrial precursor proteins occurs via contact sites. The import of the β-subunit of the F_1-ATPase ($F_1\beta$) and subunits IV (cytochrome c_1) and V (FeS protein) of the bc_1-complex were performed at low temperatures [15,25]. This yielded translocation intermediates that spanned both mitochondrial membranes. The amino terminal presequence was cleaved off by the processing peptidase in the mitochondrial matrix, whereas a carboxy terminal part of the precursor was still accessible to externally added proteases. Only the first import step into translocation contact sites required $\Delta\Psi$. Raising the temperature led to complete import of the precursors, even in the absence of $\Delta\Psi$. Similar translocation intermediates were obtained when specific antibodies were prebound to the precursors before exposing them to mitochondria. The amino terminus was imported into mitochondria and proteolytically processed, while the carboxy terminal part was kept outside of the outer membrane by the bound antibody.

Precursors without amino terminal presequences, like the precursor to the ADP/ATP carrier, were also shown to be imported via translocation contact sites. Precursor to the ADP/ATP carrier was incubated with mitochondria at a low temperature in the presence of a membrane potential. The precursor, which associated with mitochondria, was still accessible to proteases from the outside. By raising the temperature, however, the completion of import into the inner membrane could be performed in the absence of $\Delta\Psi$. Thus the $\Delta\Psi$-dependent interaction of precursor with the inner membrane had already taken place when precursor was incubated with mitochondria at a low temperature. The precursor therefore extended from the outside of the outer membrane into the inner membrane, thereby spanning translocation contact sites [16].

The antibody-bound translocation intermediate of $F_1\beta$ in contact sites was labeled with protein-A gold particles. Morphometric and statistical evaluation of electron micrographs showed that most of the gold particles were located at sites of close contact between outer and inner membranes. Thus protein import seems to occur at the morphological contact sites [26].

When most of the outer membrane was removed by treating mitochondria with digitonin, the contact sites still retained a part of the outer membrane. These ''mitoplasts'' showed the normal rate in import of $F_1\beta$. Since the presence of additional import sites on the exposed inner membrane could be excluded, this suggests that practically all import of $F_1\beta$ occurs at translocation contact sites. Furthermore, by separating submitochondrial vesicles on a sucrose gradient, the translocation contact sites could be enriched 10-fold [26].

ENERGY REQUIREMENT OF IMPORT

It has been known for many years that mitochondrial protein import requires energy [4,27]. The transport of proteins into or across the inner membrane depends on an energized inner membrane [12,28,30]. The necessary energy form is the electrical component $\Delta\Psi$ of the total protonmotive force [42]. Only the initial interaction of positively charged precursor domains with the inner membrane seems to require $\Delta\Psi$ [16,25]. It is suggested that the role of $\Delta\Psi$ involves an electrophoretic effect on positively charged precursor domains.

Recently, it was shown that the import of mitochondrial precursor proteins also needs nucleoside triphosphates (NTPs), independently of the requirement for a membrane potential. This was demonstrated for the import of $F_1\beta$ [31] ADP/ATP carrier, fusion proteins between F_0-ATPase subunit 9 and dihydrofolate reductase [32], and porin [33]. All import steps that are involved in the transport of precursors from the cytosol into the outer membrane required NTPs. The insertion of precursors into the outer membrane required higher levels of NTPs than the binding to receptor sites on the mitochondrial surface. Three precursors, each of which contained the presequence of F_0-ATPase subunit 9, but different carboxy terminal parts, required different levels of NTPs for import. The protease resistance of in vitro synthesized precursor proteins in reticulocyte lysate was decreased by the presence of NTPs. We conclude that NTPs affect the cytosolic conformation of precursor proteins. We propose that cytosolic proteins act together with NTPs to keep precursor proteins in an import-competent conformation [32].

PROTEOLYTIC PROCESSING

The amino terminal presequences are proteolytically removed by the processing peptidase, which is located in the mitochondrial matrix [34–36]. The processing peptidase from *Neurospora crassa* has been enriched 2,000-fold over mitochondria [37]. Proteolytic processing is not a prerequisite for transport across the mitochondrial membranes, since precursors can be imported into the matrix in the absence of processing [15,38].

For precursors that are proteolytically processed in two steps, the first processing step is always performed by the matrix-located processing peptidase. The second processing step is either done by the processing peptidase as well [36] or by different processing activities that are assumed to be located in the intermembrane space [29].

INTRAMITOCHONDRIAL SORTING OF PRECURSORS

Whereas the mature FeS protein of the bc_1-complex is located on the outer face of the inner membrane, it has been shown that the precursor is first completely transported into the matrix via translocation contact sites. After proteolytic processing by the processing peptidase, the FeS protein is redirected back across the inner membrane [15]. Proteolytic processing takes place in two steps, but it is not known so far whether the second processing event occurs before, during, or after retranslocation. The insertion of the FeS cluster into the protein seems to occur in the matrix space [Hartl and Neupert, unpublished]. In agreement with the endosymbiotic hypothesis of mitochondrial origin, it is proposed that folding and assembly pathways have been conserved during evolution. After transfer of mitochondrial genes to the nucleus, presequences and import via contact sites have been introduced to transport precursors back into the mitochondrial matrix—the equivalent of the procaryotic cytoplasm. From there, the conserved folding and assembly pathways, which were already present in the procaryotic ancestor, are used.

A different import pathway for intermembrane space proteins is proposed in the following model. Precursors to intermembrane space proteins are kept in the inner membrane by the second part of the presequence, called the "stop transfer signal" [39]. The first part of the presequence could then be cleaved off by the processing peptidase, whereas the second cleavage step is performed by a protease in the intermembrane space. In this model, the precursors should diffuse laterally from contact sites into the intermembrane space.

The transport of cytochrome *c* into the intermembrane space is an exception in many respects. The import is independent of translocation contact sites and of the membrane potential. Apocytochrome *c*, which does not contain a cleavable presequence, is translocated across the outer membrane. Holocytochrome *c* is formed by covalent attachment of heme in the intermembrane space [40]. Apocytochrome *c* from a *Neurospora* mutant in which only the carboxyterminus is altered is imported into mitochondria at a very low efficiency [41]. Cytochrome *c* seems to circumvent the "complicated" import pathway of other intermembrane space proteins that carry cleavable amino terminal signal sequences and are imported via translocation contact sites in a $\Delta\Psi$-dependent manner.

In summary, it might well be that several (at least three) different pathways exist for the import of proteins into the mitochondrial intermembrane space.

PERSPECTIVES

During the last few years, functional studies have resolved the pathways of mitochondrial protein import into several distinct steps. Now the most important aims are the isolation and characterization of the components involved in protein import. These include the precursor proteins, cytosolic cofactors, receptors, components of contact sites, and other components of the translocation apparatus, the processing peptidase of the matrix and other processing enzymes, putative cofactors inside mitochondria, components of the transport apparatus for retranslocation of proteins across the inner membrane, and components required for assembly of subunits into protein complexes. Thus it should be possible to understand mitochondrial protein import at a molecular level.

ACKNOWLEDGMENTS

We thank Dr. D.W. Nicholson for critically reading the manuscript.

REFERENCES

1. Hay R, Böhni P, Gasser S: Biochim Biophys Acta 779:65–87, 1984.
2. Harmey MA, Neupert W: In Martonosi A (ed): "The Enzymes of Biological Membranes," Vol. 4. New York: Plenum Publishing Co., 1985, pp 431–464.
3. Pfanner N, Neupert W: In Lee CP (ed): "Current Topics in Bioenergetics," Vol. 15. New York: Academic Press, 1987, pp 177–219.
4. Hallermayer G, Neupert W: In Bücher T, Neupert W, Sebald W, Werner S (eds): "Genetics and Biogenesis of Chloroplasts and Mitochondria." Amsterdam: North-Holland, 1976, pp 807–812.
5. Hallermayer G, Zimmermann R, Neupert W: Eur J Biochem 81:523–532, 1977.
6. Harmey MA, Hallermayer G, Korb H, Neupert W: Eur J Biochem 81:533–544, 1977.
7. Zimmermann R, Neupert W: Eur J Biochem 109:217–229, 1980.
8. Hurt EC, Pesold-Hurt B, Schatz G: FEBS Lett 178:306–310, 1984.
9. Horwich AL, Kalousek F, Mellman I, Rosenberg LE: EMBO J 4:1129–1135, 1985.
10. van Loon APGM, Young ET: EMBO J 5:161–165, 1986.
11. Hennig B, Neupert W: Eur J Biochem 121:203–212, 1981.
12. Gasser SM, Daum G, Schatz G: J Biol Chem 257:13034–13041, 1982.
13. Zwizinski C, Schleyer M, Neupert W: J Biol Chem 259:7850–7856, 1984.
14. Pfaller R, Freitag H, Harmey MA, Benz R, Neupert W: J Biol Chem 260:8188–8193, 1985.

15. Hartl F-U, Schmidt B, Wachter E, Weiss H, Neupert W: Cell 47:939–951, 1986.
16. Pfanner N, Neupert W: J Biol Chem 262:7528–7536, 1987.
17. Pfaller R, Neupert W: EMBO J 6:2635–2642, 1987.
18. Hennig B, Köhler H, Neupert W: Proc Natl Acad Sci USA 80:4963–4967, 1983.
19. Schmidt B, Pfaller R, Pfanner N, Schleyer M, Neupert W: In Quagliariello E, et al (eds): ''Achievements and Perspectives of Mitochondrial Research,'' Vol II. Amsterdam: Elsevier Science Publishers B.V., 1985, pp 389–396.
20. Zwizinski C, Schleyer M, Neupert W: J Biol Chem 258:4071–4074, 1983.
21. Riezman H, Hay R, Witte C, Nelson N, Schatz G: EMBO J 2:1113–1118, 1983.
22. Zimmermann R, Hennig B, Neupert W: Eur J Biochem 116:455–460, 1981.
23. Teintze M, Slaughter M, Weiss H, Neupert W: J Biol Chem 257:10364–10371, 1982.
24. Hackenbrock CR: Proc Natl Acad Sci USA 61:589–602, 1968.
25. Schleyer M, Neupert W: Cell 43:339–350, 1985.
26. Schwaiger M, Herzog V, Neupert W: J Cell Biol 105:235–246, 1987.
27. Nelson N, Schatz G: Proc Natl Acad Sci USA 76:4365–4369, 1979.
28. Schleyer M, Schmidt B, Neupert W: Eur J Biochem 125:109–116, 1982.
29. Gasser SM, Ohashi A, Daum G, Böhni PC, Gibson J, Reid GA, Yonetani T, Schatz G: Proc Natl Acad Sci USA 79:267–271, 1982.
30. Kolanski DM, Conboy JG, Fenton WA, Rosenberg LE: J Biol Chem 257:8467–8471, 1982.
31. Pfanner N, Neupert W: FEBS Lett 209:152–156, 1986.
32. Pfanner N, Tropschug M, Neupert W: Cell 49:815–823, 1987.
33. Kleene R, Pfanner N, Pfaller R, Link TA, Sebald W, Neupert W, Tropschug M: EMBO J 6:2627–2633, 1987.
34. Böhni P, Gasser S, Leaver C, Schatz G: In Kroon AM, Saccone C (eds): ''The Organization and Expression of the Mitochondrial Genome.'' Amsterdam: Elsevier/North-Holland, 1980, pp 423–433.
35. McAda PC, Douglas MG: J Biol Chem 257:3177–3182, 1982.
36. Schmidt B, Wachter E, Sebald W, Neupert W: Eur J Biochem 144:581–588, 1984.
37. Hawlitschek G, Schneider H, Schmidt B, Tropschug M, Hartl F-U, Neupert W: 1988, submitted.
38. Zwizinski C, Neupert W: J Biol Chem 258:13340–13346, 1983.
39. Hurt EC, van Loon APGM: Trends Biochem Sci 11:204–207, 1986.
40. Nicholson DW, Köhler H, Neupert W: Eur J Biochem 164:147–157, 1987.
41. Stuart RA, Neupert W, Tropschug M: EMBO J 6:2131–2137, 1987.
42. Pfanner N, Neupert W: EMBO J 4:2819–2825, 1985.

Molecular Biology of Intracellular Protein
Sorting and Organelle Assembly 133–138 (1988)

Discussion Summary: Prokaryotic Secretion

Thomas J. Silhavy and William T. Wickner

Molecular Biology Department, Princeton University, Princeton, New Jersey 08544 (T.J.S.); and Department of Biological Chemistry, UCLA School of Medicine and Molecular Biology Institute, Los Angeles, California 90024 (W.T.W.)

Although protein localization in bacteria was not a major focus of this symposium, it is clear that many basic features of the export process have been conserved throughout biology. It is also clear that investigators working with bacteria are encountering the same sorts of problems that plague scientists working with yeast or higher eukaryotes and that they are entertaining similar solutions and concepts. This realization, coupled with the fact that protein secretion in bacteria is a topic of considerable industrial concern, swelled the ranks of attendees and this, in turn, stimulated productive discussion.

The topics discussed in this workshop fell into five major areas, listed below.

WHAT IS A SIGNAL SEQUENCE?

The importance of the signal or leader sequence for protein export has been recognized for more than a decade. However, the function(s) performed by this sequence during the export process remains largely unknown. This functional mystery has not been resolved by the compilation and analysis of the amino acid sequence of large numbers of different signal sequences [1]. Such studies have revealed a surprising lack of homology, and it is hard to understand how such diverse sequences could possibly perform analogous roles. Recently this paradox has been made more graphic by reports that a substantial percentage of seemingly random sequences can function, at some level, to replace a signal sequence [2].

One systematic approach to the study of signal sequences involves the identification and biochemical characterization of signal sequence mutations. Such studies have been done with several different experimental systems, including both periplasmic and outer membrane proteins [for review see 3]. It is simply not possible to summarize these mutant data and present a clear picture of signal sequence function. Even if the analysis is confined to a particular gene, meaningful comment at anything but a cursory level is difficult and dangerous.

Received April 17, 1987.

It is generally agreed that a signal sequence contains three distinct domains or regions. Starting at the amino terminus, these domains include: 1) a hydrophilic segment of variable length that terminates in one or two basic amino acids; 2) a hydrophobic core containing at least eight uncharged residues; and 3) a processing site, the site of leader peptidase action.

Region I

With the exception of the basic residues, this region probably contains no export-related information, but rather is required for efficient translation initiation. Indeed, large sequences that encode normally cytoplasmic proteins can be placed here with little or no effect on export [4,5]. Mutations that substitute neutral or negatively charged amino acids for the basic residues can have a detrimental effect on export, or they can cause a decrease in overall protein synthesis [4,6]. In at least one case, this decrease in synthesis is the result of a defect in translation elongation [6]. The properties of this mutation may reflect a cellular mechanism that serves to prevent protein synthesis in the absence of translocation.

Region II

Most strongly export-defective mutations alter this region of the signal sequence, either by substituting a charged residue or by shortening the region by deletion mutation. Clearly the hydrophobic character of the signal sequence is important. However, not every charge substitution causes a strong export defect. Certain positions in the hydrophobic core appear to be more critical than others [3,7]. Secondary structure of this region is important as well. Genetic and biophysical data show that this region must form a alpha-helical structure at some point during the export process [8,9].

Region III

This region of the signal sequence is much less hydrophobic than the core; however, charged residues are rarely present. The recognition sequence for leader peptidase and the cleavage site terminate this region. A reasonable, but rather poorly conserved, consensus sequence for the cleavage site is Ala-X-Ala. Mutations that prevent proteolytic processing have been identified in several different systems. In general, they do not appear to prevent translocation of the molecule from the cytoplasm. The lack of processing may or may not, however, interfere with subsequent intracellular routing [10,11].

These rather limited insights into the critical nature of signal sequences provide no real clues concerning the mechanism of signal function. Thus a key element for understanding protein export is missing. Further progress in this area may require more careful definition of function and more specific biochemical assays. It may be that the signal sequence performs several different functions during the export process and that these functions are mediated by different, but perhaps overlapping, portions of the signal sequence. If this could be demonstrated, then the task of classifying, and thus interpreting, signal sequence mutations could be greatly simplified.

Recent studies summarized in the workshop provide an example of how the different functions of signal sequences might be subdivided. Synthetic peptides that correspond to a signal sequence have been synthesized and studied using a variety of biophysical methods. The studies directly show the importance of secondary structure, and they show that these sequences can indeed interact with lipid membranes. It seems probable that membrane interaction is one of the functions of signal sequences; this view is supported by the observation that certain signal sequence alterations that prevent export in vivo also prevent membrane interaction in vitro [12]. The question becomes, do all export-defective signal sequence mutations prevent membrane interaction? Can the portion of the signal sequence required for membrane interaction be defined? If so, this could provide an important first step towards meaningful, functional subdivision.

IS A SIGNAL SEQUENCE ENOUGH?

Many people view the signal sequence as an entity unto itself, i.e., they assume that the export information required to direct protein translocation from the cytoplasm is contained entirely within the peptide that is cleaved proteolytically from the precursor form of the exported protein. Although tidy, this belief appears incorrect.

Studies with two different inner membrane proteins demonstrate the participation of mature sequences in the process of membrane insertion. In the case of the coat protein of M13, a protein capable of self-catalyzed membrane assembly, a region within mature sequences that appears to act in concert with the leader sequence to cause membrane insertion has been identified [13]. This region is hydrophobic and has been termed a "hydrophobic helper" sequence. In addition, mutations that alter the acidic portion of the mature coat protein [14] and the basic carboxy terminus [15] have been shown to affect membrane assembly as well.

Leader peptidase, an inner membrane protein with an uncleaved signal sequence, also has hydrophobic helper domains and features of its polar carboxy terminus that are important for membrane insertion [16]. However, unlike the coat protein, leader peptidase does not assemble spontaneously. Correct membrane insertion requires components of the cellular export machinery [17]. The mode of action of these mature sequences in membrane assembly is yet unclear.

In the case of LamB, a major outer membrane protein, studies with *lacZ* gene fusions and analysis of a large collection of internal, in-frame deletions reveal that the first 28 amino acids of the protein are required, in addition to the signal sequence, for rapid and efficient export from the cytoplasm. The signal sequence alone directs translocation at a greatly reduced rate and efficiency. Here again, the role of the mature sequences is unclear [18]. These mature sequences may, together with the signal sequence, form a structural domain that facilitates signal recognition. Alternatively, they may serve as a "spacer" to prevent downstream sequences from interfering with signal sequence conformation or accessibility.

An issue related to (and often confused with) the presence of export information within mature sequences is the obvious incompatibility of certain protein sequences with the export machinery. Certain cytoplasmic enzymes can apparently not be exported. This is not because of a lack of export information as in the coat protein, leader peptidase, and LamB examples described above, but rather it is the result of sequences within the cytoplasmic enzyme that act in a pronounced negative fashion relative to export. The most thoroughly characterized example of this negative information is β-galactosidase. If a large, functional, carboxy terminal portion of this enzyme is fused to sequences of a noncytoplasmic protein, high-level synthesis of the hybrid molecule can lead to cell death. It is thought that this is the result of a lethal jamming of the export machinery, since the precursor forms of many different exported proteins can be detected in the cytoplasm of the moribund cells [3]. One explanation for these results is that the export machinery exhibits substrate specificity, and it can distinguish one protein sequence from another. No compelling molecular explanation can yet be offered for this incompatibility, and the nature of these negatively acting sequences has not been elucidated.

COMPLEXITY OF BACTERIAL PROTEIN EXPORT

The components known to be required for protein translocation across the plasma membrane of *E. coli* are the precursor proteins themselves, the lipid bilayer, ATP, and the membrane electrochemical potential. This is a very short and obviously incomplete list, and both genetic and biochemical studies indicate a much greater level of complexity.

In general, three different genetic approaches have been employed in an effort to identify and characterize components of the cellular export machinery. The first approach was to exploit the null phenotype conferred by certain signal sequence mutations to search for second-site suppressor mutations. Presumably this would help identify components of the machinery that interact directly with the signal sequence. At least four such suppressors have been identified; only one however, has been characterized in detail [20]. This suppressor, *prlA*, is discussed further below.

A second approach involves the search for mutants in which protein export is generally defective. This can be done either by exploiting certain phenotypes exhibited by *lacZ* fusion strains, or by more brute force techniques. Four separate loci have been implicated by these approaches. One, *secA*, has been studied in some detail. Another, *secY*, is an allele of the previously described *prlA* gene. Conditionally lethal mutations in either of these genes cause a general accumulation of the precursor forms of a variety of exported proteins [20,21].

A final genetic approach that has been applied for the identification of genes whose products participate in the process of protein export is to search for conditionally lethal suppressors of temperature-sensitive alleles of either *secA* or *prlA* (*secY*) [22–24]. In one case, a suppressor of *secA* was identified that mapped to *prlA* [24]. This circular interconnection is comforting and indicates the basic utility of this method. However, the other suppressors appear to act indirectly simply by slowing the growth rate. Since this effect can be mimicked by sublethal concentrations of antibiotics [25], the value of continued effort might be questioned.

Biochemical approaches for characterizing the cellular export machinery have been greatly stimulated by the development of an in vitro protein translocation system [26]. However, attempts to subfractionate either the soluble or the membrane components required for translocation have been somewhat frustrating. Furthermore, extracts prepared from the thermosensitive, export-defective mutants described above are not themselves thermosensitive, i.e., these mutants do not exhibit a rapid loss of function. Clearly protein components in both the soluble and membrane fractions are required, but their identity remains elusive [27]. No doubt some of these are involved in energy coupling, and further progress along these lines appears imminent.

ENERGY REQUIREMENTS

The requirement for a membrane electrochemical potential for translocation in vivo has been well established [28–32]. A membrane potential is also needed in vitro for optimal translocation rates and extents, although one-tenth of the translocation reaction can occur in the absence of a potential [33,34]. Similarly, ATP is absolutely required for translocation in vitro [35], yet translocation in the intact cell has not yet shown sensitivity to arsenate or other agents that lower the cellular pools of high-energy phosphate. Nevertheless, the current evidence favors the participation of two forms of energy in the translocation events.

While energy is essential for membrane assembly, the proteins that mediate this interaction are not yet identified. It is clear that the ATP must be hydrolyzed to support protein translocation across the plasma membrane, and we presume that a protein component is required to couple this hydrolysis to translocation. One (currently fashionable) postulate is that ATP might be needed to drive the unfolding or refolding of preproteins prior to their translocation. Several lines of evidence suggest that the requirement for the membrane electrochemical potential may be, at least in part, direct: 1) M13 coat protein requires the membrane potential for membrane assembly, even though its assembly is unusually simple in terms of required protein catalysts; 2) a point mutation in the gene for coat protein bypasses the requirement for the potential for membrane insertion [14]; and 3) different proteins have different levels of requirement for the potential for the translocation reaction [31].

Attempts to follow the fate of the gamma-phosphoryl group from ATP during protein translocation have been frustrated by the high levels of other cellular ATP hydrolases. Recently there has been progress in this regard. The soluble components of an in vitro protein synthesis reaction can be fractionated. It appears that the precursor protein and an additional "trigger factor," with a role in the correct folding of pro-OmpA, are all that is required [36]. Isolation of these proteins will remove the bulk of the soluble ATPases from the system. In addition, membrane vesicles are now available from strains that genetically lack the major F1-ATP synthase. Hopefully, this will now permit us to follow the fate of the gamma-phosphoryl group as it is coupled to preprotein translocation.

SYNERGY OF BIOCHEMISTRY AND GENETICS

To establish beyond reasonable question the components of bacterial protein export and the mechanisms by which they support translocation, each must be isolated in pure form and reconstituted into an in vitro translocation reaction. Mutants must be isolated and their phenotypes established, the structural gene for each identified, and the mutant proteins characterized in vitro. Only when this combined genetic and biochemical approach gives a coherent picture will we have confidence as to the export mechanism.

Two limitations to current work have, together, prevented this synergy from being achieved. One is the inability to solubilize assembly-competent inner membrane vesicles in detergent and then remove the detergent and reconstitute competent membranes. If this reconstitution could be achieved, it would permit fractionation of the detergent extract and isolation of the relevant components for preprotein translocation. Of course, this limitation is also precisely the same for studies of protein transit into the endoplasmic reticulum or mitochondria in eukaryotic cells. The other limit is that none of the available mutants that are generally export-defective show a rapid shut-off of translocation when shifted to the nonpermissive condition. New strategies must be devised for mutant selection. Hopefully, rapid shut-off mutants would have in vitro defects that could lead to a complementation assay for protein translocation.

REFERENCES

1. Watson MEE: Nucleic Acids Res 12:5145, 1984.
2. Kaiser CA, Preuss D, Grisafi P, Botstein D: Science 235:312, 1987.
3. Benson SA, Hall MN, Silhavy TJ: Annu Rev Biochem 54:101, 1985.
4. Pollitt S, Inouye S, Inouye M: Microbiology 238, 1986.
5. Kuhn A, personal communication.
6. Hall MN, Gabay J, Schwartz M: EMBO J 2:15, 1983.
7. Stader J, Benson SA, Silhavy TJ: J Biol Chem 261:15075, 1986.
8. Emr SD, Silhavy TJ: Proc Natl Acad Sci USA 80:4599, 1983.
9. Briggs MS Gierasch LM: Biochemistry 23:3111, 1984.
10. Dalbey RE, Wickner W: J Biol Chem 260:15925, 1985.
11. Fikes JD, Bassford PJ: J Bacteriol 169:2352, 1987.
12. Briggs MS, Gierasch LM, Zlotnick A, Lear JD, DeGrado WF: Science 228:1096, 1985.
13. Kuhn A, Kreil G, Wickner W: EMBO J 5:3681, 1986.
14. Zimmermann R, Watts C, Wickner W: J Biol Chem 257:6529, 1982.
15. Kuhn A, Wickner W, Kreil G: Nature 322:335, 1986.
16. Dalbey RE, Wickner W: J Biol Chem 261:13844, 1986.
17. Wolfe PB, Wickner W: Cell 36:1067, 1984.
18. Rasmussen BA, Silhavy TJ: Genes Dev 1:185, 1987.
19. Shultz J, Silhavy TJ, Berman ML, Fiil N, Emr SD: Cell 31:227, 1982.
20. Oliver DB, Beckwith J: Cell 25:765, 1981.
21. Shiba K, Ito K, Yura T, Cerretti DP: EMBO J 3:631, 1984.
22. Oliver D: J Bacteriol 166:878, 1985.
23. Shiba K, Ito K, Yura T: J Bacteriol 166:849, 1986.
24. Brickman ER, Oliver DB, Garwin JL, Kumamoto C, Beckwith J: Molec gen Genet 196:24, 1984.
25. Lee CA, Beckwith J: J Bacteriol 166:878, 1986.
26. Chen L, Rhoads D, Tai PC: J Bacteriol 161:973, 1985.
27. Muller M, Blobel G: Proc Natl Acad Sci USA 81:7421, 1984.
28. Date T, Zwizinski C, Ludmerer S, Wickner W: Proc Natl Acad Sci USA 77:827, 1980.
29. Date T, Goodman J, Wickner W: Proc Natl Acad Sci USA 77:4669, 1980.
30. Enequist HG, Hirst TR, Harayama S, Hardy SJS, Randall LL: Eur J Biochem 116:227, 1981.
31. Daniels CJ, Bole DG, Quay SC, Oxender DL: Proc Natl Acad Sci USA 78:5396, 1981.
32. Zimmermann R, Wickner W: J Biol Chem 258:3920, 1983.
33. Geller BL, Movva NR, Wickner W: Proc Natl Acad Sci USA 83:4219, 1986.
34. Chen L, Tai P-C: J Bacteriol 168:828, 1986.
35. Chen L, Tai P-C: Proc Natl Acad Sci USA 82:4384, 1985.
36. Crooke E, Wickner W: Proc Natl Acad Sci USA 84:5216, 1987.

Molecular Biology of Intracellular Protein
Sorting and Organelle Assembly 139–145 (1988)

How Are Peroxisomal Proteins Directed to and Imported Inside Peroxisomes?

Gillian M. Small, Tsuneo Imanaka, and Paul B. Lazarow

The Rockefeller University, New York, New York 10021

Newly synthesized peroxisomal proteins enter pre-existing peroxisomes post-translationally in vivo. Most of these proteins are synthesized at their mature size, without cleavable targeting peptides. We have reproduced this import in vitro, using peroxisomes isolated from rat liver and from the yeast *Candida tropicalis*. This has enabled us to study the requirements and nature of the targeting information for post-translational import.

Key words: peroxisome, post-translational import, topogenic signals

It is now established that the development of peroxisomes requires the post-translational import of proteins that are synthesized on cytosolic ribosomes (Fig. 1) [1]. Many details are known about the targeting and translocation of proteins into mitochondria [2], chloroplasts [3], and nuclei [4], but the reactions involved in the import of proteins into peroxisomes are largely unknown. Many mitochondrial and chloroplast proteins are synthesized as precursor forms with a cleavable amino terminal extension that contains the targeting information. Most peroxisome proteins studied to date are synthesized at their mature size and lack cleavable topogenic sequences; this raises intriguing questions regarding their import. What is the nature and location of the targeting information? What is the process (in the absence of cleavage from a precursor to a mature form) by which the mechanism is made irreversible? These questions can be addressed by reconstituting in vitro the post-translational import of proteins into peroxisomes. With such a system it is also possible to investigate in detail the mechanism and specific requirements for the translocation step.

These questions are under study in our laboratory with two types of peroxisomes: 1) with rat liver peroxisomes we are investigating the mechanism and requirements for the occurrence of post-translational import; and 2) with peroxisomes from the yeast *Candida tropicalis*, for which we have constructed a cDNA library [5], we are investigating the topogenic sequences involved in targeting peroxisomal proteins to peroxisomes.

Received May 4, 1987.

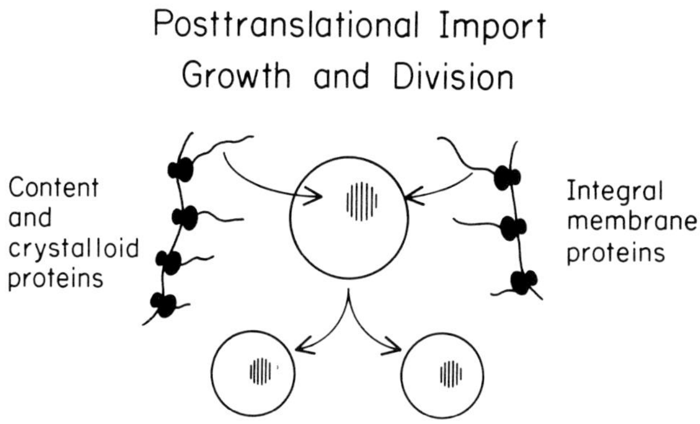

Fig. 1. Current model of peroxisome biogenesis [from reference 1].

POST-TRANSLATIONAL IMPORT

The post-translational import of acyl-CoA oxidase (the first enzyme in the peroxisomal β-oxidation system) and catalase was first accomplished in vitro with rat liver peroxisomes [6]. Import was demonstrated by cosedimentation of cell-free translation products with peroxisomes in a sucrose gradient and by resistance of these products to proteolysis. We have now improved the efficiency of this import reconstitution to the point that 30% of added [^{35}S]acyl-CoA oxidase (AOx) is imported into highly purified peroxisomes in 30 minutes at 26°C. We have recently obtained evidence that a protein or proteins in the peroxisomal membrane is/are necessary for AOx import. Furthermore, we have found that import of this protein into rat liver peroxisomes requires the hydrolysis of ATP, but does not require a membrane potential [7]. Using pulse-chase experiments, Bellion and Goodman observed that proton ionophores prevented the import of alcohol oxidase into peroxisomes in *Candida boidinii* spheroplasts [8]. This could be due to a requirement for a membrane potential, or, alternatively, the authors suggest that lack of import could be a result of the drop in intracellular ATP concentrations that the ionophores cause. Our results suggest that the energy requirement for translocation of proteins into peroxisomes is similar to that for the import of proteins into chloroplasts, where ATP hydrolysis alone has been shown to supply sufficient energy for import [9,10]. In contrast, both ATP and an energized inner mitochondrial membrane are necessary for the import of proteins into mitochondria [11]. The function of ATP in these import systems has yet to be ascertained. We have suggested the possibility that a cyclical phosphorylation/dephosphorylation of a membrane protein might be involved [7].

TOPOGENIC SEQUENCES

We have used the yeast *C. tropicalis* to investigate the nature and location of the topogenic information that functions to target proteins to peroxisomes. Peroxisomes can be strongly induced by growth of this yeast on alkanes or oleic acid. Similarly induced are ~ 12 mRNAs, most of which encode peroxisomal proteins, including catalase, acyl-CoA oxidase, and the trifunctional protein hydratase-dehydrogenase-epimerase. Each of

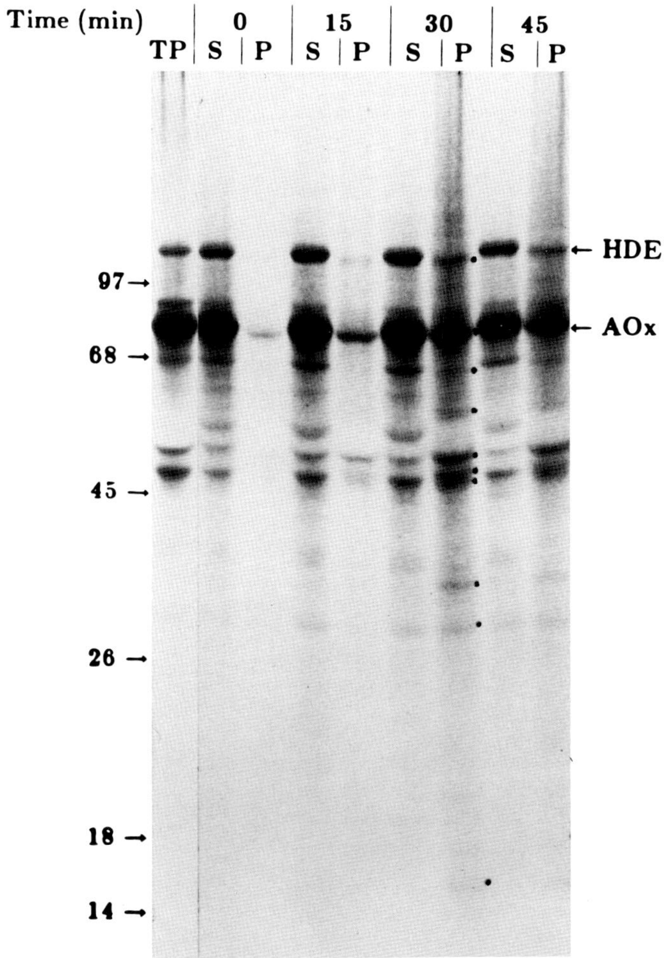

Fig. 2. Time-dependent post-translational association of peroxisomal translation products with peroxisomes. Translation products (25 μl) were incubated with peroxisomes (~ 35 μg of protein) at 26°C for the times indicated, and the peroxisomes (P) were separated from the supernatants (S) by centrifugation. Peroxisomal translation products were immunoprecipitated from equivalent amounts of supernatants and pellets (12.5% of each) and analyzed by SDS-PAGE and fluorography. TP, immunoprecipitation of an equivalent amount of translation products; HDE, trifunctional protein catalyzing three reactions in fatty acid metabolism: enoyl-CoA hydratase, β-Hydroxyacyl-CoA dehydrogenase, and β-Hydroxyacyl-CoA epimerase. AOx, acyl-CoA oxidase. Numbers with arrows indicate molecular mass markers (kD).

these is synthesized at its final size [12]. We have established an in vitro system to assay the import of cell-free translation products into purified, stable peroxisomes from *C. tropicalis* [13]. We found a specific association of ~ 10 cell-free translation products with yeast peroxisomes and have confirmed the peroxisomal nature of these proteins by immunoprecipitation with an antiserum specific for peroxisomal polypeptides (Fig. 2). This association was temperature-dependent and organelle-specific (Fig. 3). The most abundant in vitro product from the translation of oleate-induced mRNA is AOx [12]. This protein associates efficiently with peroxisomes in the in vitro import assay (Fig. 4).

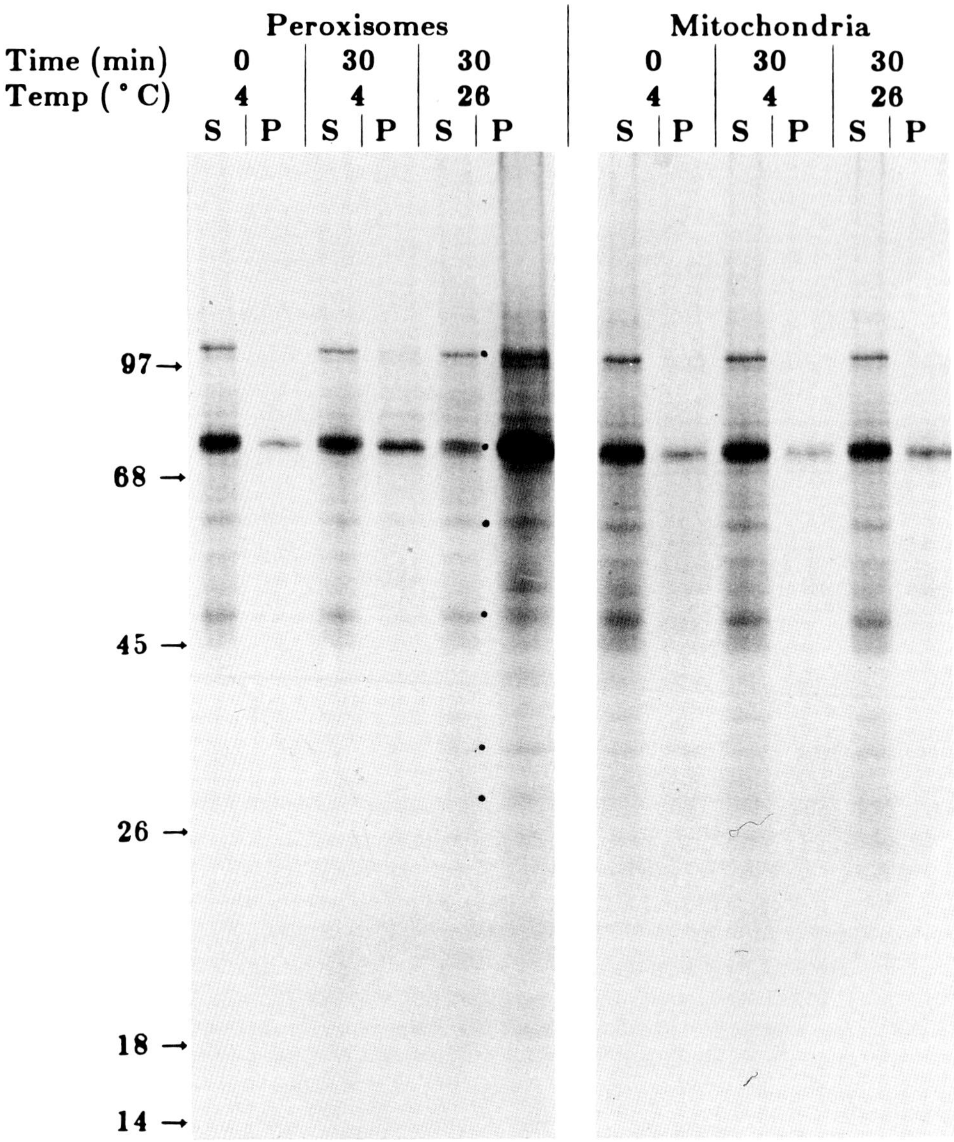

Fig. 3. Temperature dependence and organelle specificity. Incubations were done with peroxisomes (60 μg of protein) or mitochondria (64 μg of protein). Some 20% of the organelle pellets (P) and 5% of the supernatants (S) were analyzed by SDS-PAGE and fluorography. Dots indicate polypeptides that decreased in the supernatants and appeared in the pellets.

We identified a cDNA clone encoding the carboxy terminal portion of acyl-CoA oxidase [5], and we expressed this clone in vitro by subcloning it into an expression plasmid at a *Mae*1 site (Fig. 5) [14]. When transcribed and translated in vitro, this produced a 43-kD polypeptide and a smaller amount of a second polypeptide with a mass of ~ 32 kD (Fig. 6), which is consistent with the occurrence of secondary initiation at the next downstream methionine codon (Fig. 5). Both polypeptides associated with peroxisomes

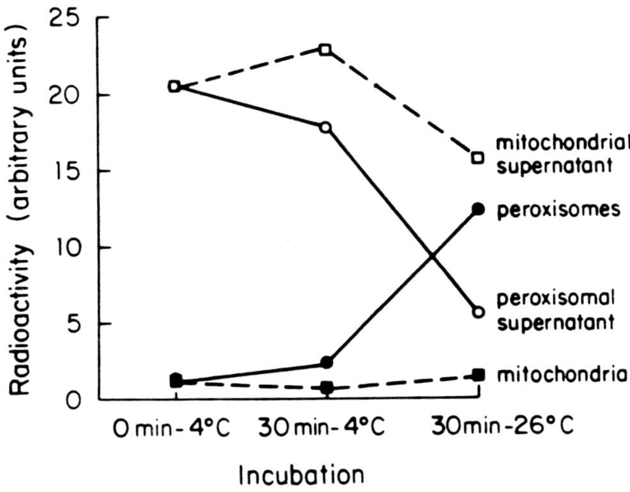

Fig. 4. Quantitation of [^{35}S]-labeled acyl-CoA oxidase in the experiment of Figure 3. Radioactivity (measured as the area under the peak in a densitometric scan) has been corrected for the four-fold difference in the amounts of supernatants and pellets analyzed by SDS-PAGE. Although the X axis does not represent a continuous function, the points have been connected for visual clarity.

in vitro, and approximately 60% of the peroxisome-associated products were protease-resistant, indicating that they were imported into the organelle (Fig. 6) [14]. These results show, unexpectedly, that there is sufficient information in the carboxy terminal portion of the protein to target it to peroxisomes.

Our findings suggest that the targeting of peroxisomal proteins differs from that of mitochondrial proteins, which rely on amino terminal (pre)sequences for both intracellular targeting and intramitochondrial sorting [15]. Targeting of nuclear proteins to the nucleus has been shown to be much more variable. The large T antigen of SV 40 has an amino terminal signal [16], while influenza virus nucleoprotein has a carboxy

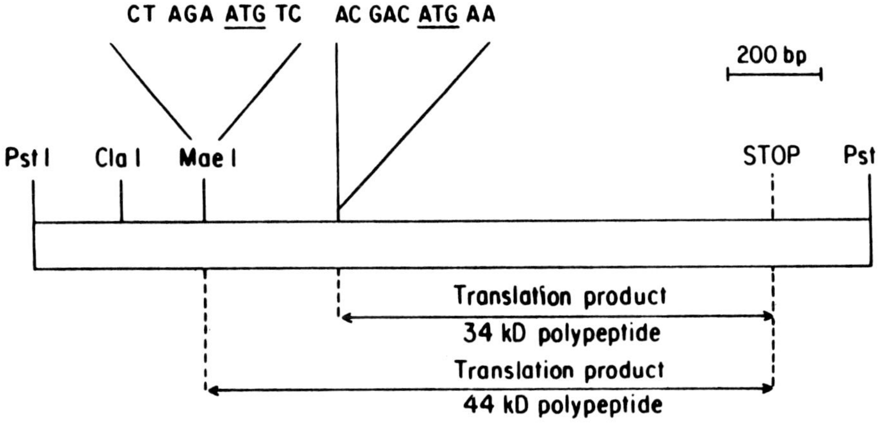

Fig. 5. Expression strategy for the acyl-CoA oxidase cDNA. The 1161-bp *Mae*1-*Pst*1 fragment was cloned into the expression vector pGEM3. The position of the two internal ATGs that could act as initiators of translation are indicated, together with the predicted sizes of the translation products.

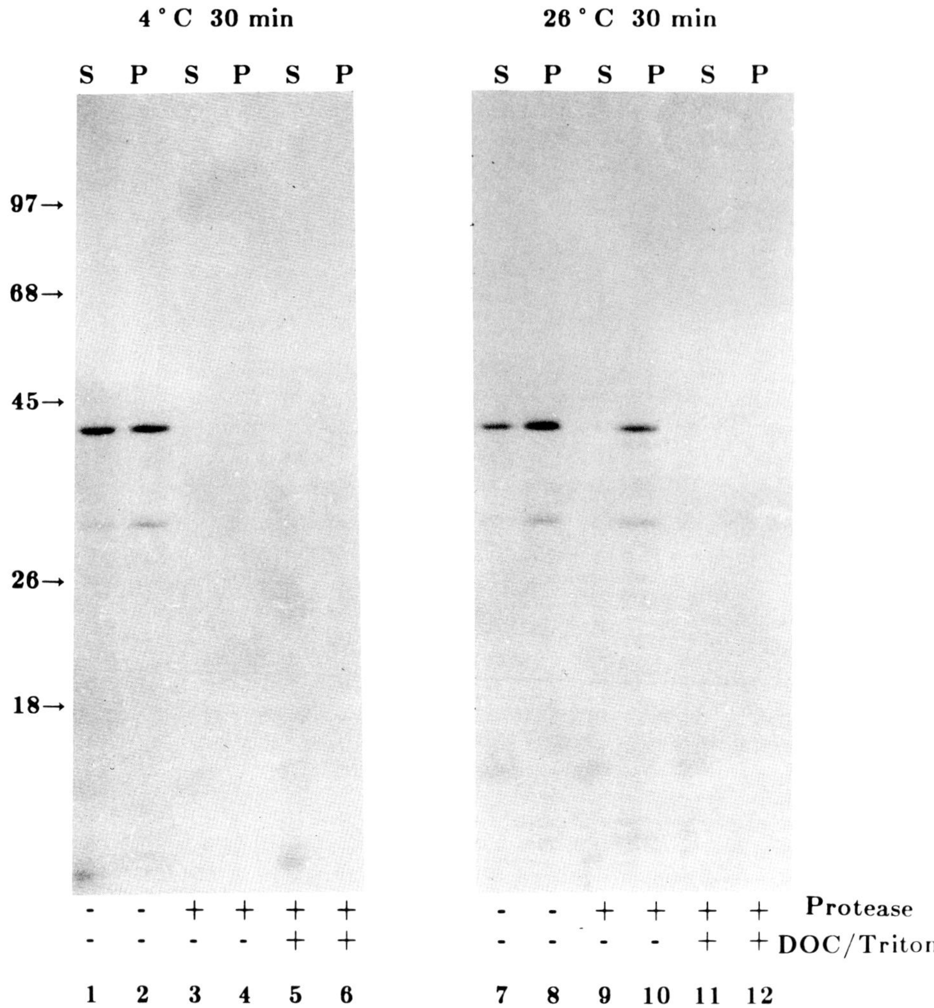

Fig. 6. Temperature-dependent import into peroxisomes of the 44- and 34-kD in vitro expression products of the AOx cDNA. Newly synthesized translation products (300 μl) were mixed with peroxisomes (2.2 mg of protein) and divided into two equal samples. Incubation was at 4°C (left) or 26°C (right) for 30 minutes. Following this, each sample was divided into three aliquots. Samples were (+) or were not (−) digested with 1 μg of thermolysin at 4°C for 30 minutes in the presence (lanes 5,6,11,12) or absence (lanes 1–4 and 7–10) of 1% sodium deoxycholate and 1% Triton X-100. Supernatants and pellets were then separated and analyzed as in Figure 3.

terminal signal [17], and yeast Mat $\alpha 2$ has an amino terminal signal and a second signal in the carboxy terminal portion of the protein [18,4]. We are currently investigating whether other portions of the acyl-CoA oxidase protein are also capable of being imported into peroxisomes, and have evidence for a second topogenic signal in the amino terminal region [19].

Clearly much is yet to be learned regarding the mechanisms involved in targeting and translocation of peroxisomal proteins into peroxisomes. With the tools now available, we are able to address these questions.

ACKNOWLEDGMENTS

This research was supported by NSF grant DCB 83-09222 and NIH grant AM 19394.

REFERENCES

1. Lazarow PB, Fujiki Y: Annu Rev Cell Biol 1:489–530, 1985.
2. Hay R, Bohni P, Gasser S: Biochim Biophys Acta 779:65–87, 1984.
3. Schmidt GW, Mishkind ML: Annu Rev Biochem 55:879–912, 1986.
4. Dingwall C, Laskey RA: Annu Rev Cell Biol 2:367–390, 1986.
5. Rachubinski RA, Fujiki Y, Lazarow PB: Proc Natl Acad Sci USA 82:3973–3977, 1985.
6. Fujiki Y, Lazarow PB: J Biol Chem 260:5603–5609, 1985.
7. Imanaka T, Small GM, Lazarow PB: J Cell Biol 105:2915–2922, 1987.
8. Bellion E, Goodman JM: Cell 48:165–173, 1987.
9. Flugge UI, Hinz G: Eur J Biochem 160:563–570, 1986.
10. Schindler C, Hracky R, Soll J: Z Naturforsch 41c:159–163, 1987.
11. Eilers M, Oppliger W, Schatz G: EMBO J 6:1073–1077, 1987.
12. Fujiki Y, Rachubinski RA, Zentella-Dehesa A, Lazarow PB: J Biol Chem 261:15787–15793, 1986.
13. Small GM, Imanaka T, Shio H, Lazarow PB: Mol Cell Biol 7:1848–1855, 1987.
14. Small GM, Lazarow PB: J Cell Biol 105:247–250, 1987.
15. Hurt EC, van Loon, A PGM: Trends Biochem Sci 11:204–207, 1986.
16. Kalderon D, Roberts BL, Richardson WD, Smith AE: Cell 39:499–509, 1984.
17. Davey J, Dimmock NJ, Colman A: Cell 40:667–675, 1985.
18. Hall MN, Hereford L, Herskowitz I: Cell 36:1057–1065, 1984.
19. Small AM, Szabo LJ, Lazarow PB: Submitted for publication.

Molecular Biology of Intracellular Protein
Sorting and Organelle Assembly 147–151 (1988)

Discussion Summary: Role of Glycosylation in the Intracellular Transport of Glycoproteins

Pamela Stanley

Department of Cell Biology, Albert Einstein College of Medicine, Bronx, New York 10461

Determining roles for carbohydrates in the secretion or localization of glycoproteins is complicated because of the covalent association of carbohydrate and protein. In the case of N-linked glycosylation, this association occurs on nascent peptide chains in the rough endoplasmic reticulum [1]. Changes in the localization of a glycoprotein that correlate with changes in its glycosylation may therefore reflect direct or indirect roles of carbohydrates, a direct effect being one in which the actual structure of the carbohydrate is important (implying the existence of a carbohydrate-binding "receptor") and an indirect effect being one in which carbohydrate is necessary for correct folding of the molecule. In the latter case, altered compartmentalization would be due primarily to effects on protein conformation, whereas in the former, it would be due to a lack of specific recognition.

In recent years, investigations of roles for carbohydrates in glycoprotein trafficking have included the use of drugs that inhibit or alter N-linked glycosylation [2], mutant cells that synthesize truncated or novel carbohydrates [3,4] and cloned glycoproteins mutated in vitro to lack specific glycosylation sites [5]. Many different glycoproteins have been investigated by these approaches, but only one group, the lysosomal enzymes, has been shown to be compartmentalized on the basis of carbohydrate recognition [6]. Lysosomal enzymes carry phosphorylated mannose residues that are recognized by the mannose-6-phosphate receptor and subsequently transported to lysosomes.

The question that provided the focus for discussion in this workshop was "Are there specific roles for carbohydrates in the trafficking of glycoproteins other than lysosomal enzymes?" Four experimental systems were discussed. Each of them used site-directed mutagenesis and deletion or addition of N-linked glycosylation sites in cloned glycoproteins to explore the effects of altered carbohydrate content on glycoprotein secretion or membrane localization.

Received June 6, 1987.

INFLUENZA HEMAGGLUTININ (Abstract E215)

The hemagglutinin (HA) from X31, the A/Aichi/68 strain of influenza virus, has been cloned into SV40 vectors and may be expressed in CV-1 cells [7]. This hemagglutinin has seven N-linked glycosylation sites. To determine if any particular carbohydrate moiety was necessary for secretion, the X31 HA was mutated in vitro to eliminate each of its seven glycosylation sites in turn by changing either the Asn or Ser/Thr residue at each Asn-X-Ser/Thr glycosylation sequence. All of the HA molecules that carried six instead of seven carbohydrate moieties were transported to the cell surface as efficiently as the parent molecule. Therefore glycosylation at a particular site appears not to be required for secretion. Elimination of all glycosylation sites by mutagenesis caused the HA to remain in the ER, as observed following tunicamycin treatment.

Interestingly, HA mutant with only one or three N-linked carbohydrate moieties were very inefficiently transported out of the ER. Thus one carbohydrate moiety was not sufficient for transit of the HA to the plasma membrane, in contrast to the situation observed for the G glycoprotein of vesicular stomatitis virus (VSV) and for a fusion protein containing rat growth hormone and the membrane anchor region of G from VSV [5,8]. Both the latter molecules required only one carbohydrate moiety for efficient plasma membrane localization, indicating the possible existence of a specific recognition step involving carbohydrate [5,8]. However, the results might equally well be explained by effects of carbohydrate on protein folding. The data from the X31 HA appear to be more consistent with that interpretation. Combined with the evidence that indicates oligomerization of HA is a prerequisite for exit from the ER [9,10], the glycosylation data would be consistent with a model in which lack of carbohydrates alters oligomerization and/or appropriate folding of HA oligomers. Evidence consistent with this model was provided by the finding that in some mutants, addition of a supernumerary glycosylation site, at a position in HA never previously found to carry carbohydrate, resulted in molecules with eight N-linked carbohydrates that were temperature-sensitive for transport (i.e., probably did not fold properly).

ROTAVIRUS VP7 (Abstract E237)

Results from deletion mutants of cloned VP7, a glycoprotein of rotavirus that is normally confined to the ER, also suggest that protein conformation rather than specific glycosylation is important in the trafficking of glycoproteins. The VP7 glycoprotein is found exclusively in the ER or mature rotavirions in infected cells as a membrane-anchored molecule with one N-linked carbohydrate moiety [11]. It has two hydrophobic domains in tandem at the N-terminus. Deletion of small sections in and past the second hydrophobic domain of VP7 resulted in secretion of the molecule into the medium [12]. The secreted VP7 was found in both a glycosylated form and a nonglycosylated form in approximately equal amounts, suggesting that, in this molecule, secretion is equally efficient in the presence and the absence of carbohydrate. An unglycosylated form that remained resident in the ER was also produced by one of the deletions. Thus it appears that for rotavirus VP7, glycosylation is not important to its targeting. Similar results have been found for a number of glycoproteins in cells treated with tunicamycin [13]. Tunicamycin has little or no affect on the membrane localization or secretion of several different types of glycoproteins [13].

HUMAN CHORIONIC GONADOTROPHIN (Abstract E233).

To determine the role of glycosylation in the secretion of human chorionic gonadotrophin (hCG), the α and β-subunits were cloned into expression vectors, and their synthesis was examined in transfected Chinese hamster ovary (CHO) cells. The α-subunit has two N-linked glycosylation sites. Deletion of the site at position 52 by in vitro mutagenesis did not alter secretion. However, deletion of the site at position 78 caused a decrease in secretion and was accompanied by an increase in degradation of the α-subunit. In contrast, the hCG β-subunit was not more rapidly degraded, even in tunicamycin-treated cells, although the half-time for secretion of hCG β under those conditions was increased significantly.

Further studies of the hCG β-subunit were performed in ldlD CHO cells, a mutant line defective in glucose-4-epimerase and therefore essentially unable to add galactose or N-acetylgalactosamine to glycoproteins [14]. By altering the culture conditions so that either N-linked glycosylation or O-linked glycosylation approached normal [14], it was possible to observe the role of both types of moiety separately on the secretion of the hCG β-subunit. Molecules that essentially lacked O-linked moieties but possessed fully glycosylated, N-linked carbohydrates were secreted efficiently. The combined data with the hCG-α and hCG-β glycoproteins is consistent with roles for glycosylation in molecular folding as well as in protection of portions of protein from proteolytic degradation.

GLUCOSE REGULATORY PROTEIN OR BiP (Abstract E121)

The final presentation concerned the role of carbohydrates in the interaction of glycoproteins with a glucose regulatory protein (GRP78) that appears to be identical to a binding protein (BiP) in the ER originally observed to bind immunoglobulin H chains in an overproducing myeloma cell [15]. The function of GPP78 or BiP, is not known. Although it has been found associated mainly with truncated or altered glycoproteins, and thus is often proposed to fulfill a "garbage-removal" function, it is likely that it has a more specific function in the transit of glycoproteins through the ER.

To examine the role of GRP78/BiP in the secretion of factor VIII, cloned factor VIII and a deletion mutant lacking 900 internal amino acids that contain 20 N-linked glycosylation sites were expressed in transfected CHO cells. By SDS gel electrophoresis of immunoprecipitates, it was found that approximately 50% of the newly synthesized, full-length factor VIII was associated with BiP in the ER, and a majority of this was never secreted. Deletion of the 20 N-linked glycosylation sites caused less of the factor VIII to be associated with BiP and all of it to be eventually secreted. This indicates that the glycosylation of factor VIII is important in its binding to BiP, either through a direct interaction with carbohydrate or, more likely, due to the effects of glycosylation on the conformation of the molecule. Consistent with this interpretation is the finding that tunicamycin treatment caused an increased retention of both the parent factor VIII and the deleted molecules.

It is interesting that apparently opposite results were obtained with cloned tissue plasmingen activator (TPA). This glycoprotein has three N-linked glycosylation sites. A glycosylation-deficient mutant obtained by site-directed mutagenesis of each site was not efficiently secreted and was found mainly in association with GRP78/BiP. However, the glycosylated TPA molecule was secreted efficiently and only transiently associated with GRP78/BiP. One interpretation of these disparate results is that GRP78/BiP associates with certain protein sequences whose accessbility in a folded protein depends on glycosylation. Deletion of the glycosylated region removes the protein as well as the carbohydrate sequences, and, in the case of factor VIII, this resulted in decreased association with GRP78/BiP. However, if only the carbohydrate moieties are removed by tunicamycin treatment or mutagenesis, association with GRP78/BiP usually increases—presumably as native conformations are not attainable.

DISCUSSION AND CONCLUSIONS

The general conclusion of this workshop was that, to date, there is no evidence for a general role for carbohydrates in the secretion or localization of glycoproteins, with the notable exception of lysosomal enzymes. Although deletion of individual glycosylation sites can have a dramatic affect on the localization of certain glycoproteins, in others it is possible to obtain normal compartmentalization even in the complete absence of carbohydrates. Most of the discussion focused on whether there might exist a receptor that recognizes a general feature of N-linked carbohydrates (e.g., the "core" region) and thereby facilitates transport from the ER to the Golgi or from the Golgi to the plasma membrane. However, the several examples consistent with a "receptor hypothesis" are also consistent with a "conformation hypothesis."

Although the consensus of the workshop was that conformation effects are probably the primary cause of altered glycoprotein trafficking when glycosylation patterns are changed, it remains a difficult point to prove. The increased association of certain underglycosylated glycoproteins with GRP78/BiP suggests a potential mechanism for retaining incorrectly folded proteins in the ER. The real function of GRP78/BiP may, of course, be in protecting proteins from intramolecular interactions while they are folding, as suggested by Pelham [15]. Although oligomerization of the influenza HA appears to correlate with exit from the ER [9], this does not seem to be the case for the G glycoprotein of VSV (J. Rose, Yale University). Thus the rules for each glycoprotein are likely to be specific and to depend on their particular conformational restraints. Searching for general rules that pertain to all glycoproteins would seem to be a difficult and possibly futile project. An example of the complexities to be faced was brought up in discussion of glycosylation mutants of the human histocompatability HLA-2 molecule [16]. Deletion mutagenesis of the Asn-X-Ser/Thr consensus sequon resulted in two mutant molecules that were not glycosylated and not expressed at the cell surface (presumably because of their inability to associate with β2-microglobulin). However, a second mutation in one of the glycosylation mutants caused it to be very efficiently expressed at the cell surface, even though it still lacked carbohydrate. The second mutation apparently corrected the altered conformation induced by the loss of carbohydrate and, in fact, appeared to compensate for the carbohydrate completely. This result clearly demonstrates the potential pitfalls in interpreting site-specific mutagenesis studies.

The overall conclusion of this session was that mutation or deletion of N-linked glycosylation sites can have a dramatic effect on the ability of a glycoprotein to transit the exocytic pathway or to interact with other proteins. However, the effects observed appear to be different for each glycoprotein. No general rules can be deduced, for the probable reason that all the observed affects can be explained by specific conformational changes. At this time there is no direct evidence to implicate carbohydrates in a receptor-mediated exocytosis pathway or in localization to ER, Golgi, or plasma membranes.

REFERENCES

1. Kornfeld R, Kornfeld S: Annu Rev Biochem 54:631, 1985.
2. Elbein AD: CRC Crit Rev Biochem 16:21, 1984.
3. Briles EB, Int Rev Cytol 75:101, 1982.
4. Stanley P: Annu Rev Genet 18:525, 1984.
5. Machamer CE, Florkiewicz RZ, Rose JK: Mol Cell Biol 5:3074, 1985.
6. Von Figura K, Hasilik A: Annu Rev Biochem 55:167, 1986.
7. Gething MJ, Sambrook J: Nature 300:598, 1982.
8. Guan JL, Machamer CE, Rose JK: Cell 42:489, 1985.
9. Gething MJ, McCammon K, Sambrook J: Cell 46:939, 1986.
10. Copeland CS, Doms RW, Bolzan EM, Webster RG, Helenius A: J Cell Biol 103:1179, 1986.
11. Kabcenell AK, Atkinson PH: J Cell Biol 101:1270, 1985.
12. Poruchynsky MS, Tyndall C, Both GW, Sato F, Bellamy AR, Atkinson PH: J Cell Biol 101:2199, 1985.
13. Olden K, Parent JB, White SL: Biochim Biophys Acta 650:209, 1982.
14. Kingsley DM, Kozarsky KF, Hobbie L, Krieger M: Cell 44:749, 1986.
15. Pelham HRB: Cell 46:959, 1986.
16. Santos-Aguado J, Biro PA, Fuhrmann U, Strominger JL, Barbosa JA: Mol Cell Biol 7:982, 1987.

Molecular Biology of Intracellular Protein Sorting and Organelle Assembly 153–160 (1988)

Plasma Membrane Protein Intermediates Are Present in the Secretory Vesicles of Yeast

Cherie L. Holcomb, William Hansen, Tina Etcheverry, and Randy Schekman

Department of Biochemistry, University of California, Berkeley, California 94720 (C.L.H., R.S.); Department of Biochemistry and Biophysics, University of California Medical School, San Francisco, California 94143 (W.H.); and Genentech, 460 Pt. San Bruno Blvd., South San Francisco, California 94080 (T.E.)

Yeast cells externalize both plasma membrane and secreted proteins in a process that is mediated by secretory vesicles that form at the Golgi complex. A procedure for purifying yeast secretory vesicles from a mutant strain (*sec1*) in which these vesicles accumulate at a restrictive temperature was devised to test the possibility that both classes of protein are transported to the cell surface in the same vesicle. Fractions obtained by this procedure are enriched at least five-fold for secretory vesicles over other cellular membrane components. The vesicles are composed of a distinct set of polypeptides but are not clathrin-coated and do not contain a newly synthesized vacuolar protein. From the latter result, it is evident that vacuolar proteins are sorted away from secretory proteins prior to formation of the vesicles that accumulate in *sec1* cells. In contrast, newly synthesized plasma membrane and secretory proteins fractionate coincidentally throughout the vesicle isolation procedure. To confirm this colocalization, secretory vesicles were complexed to *Staphylococcus aureus* cells, using an antibody directed against the major plasma membrane ATPase. The secretory enzyme, acid phosphatase, is recovered quantitatively in this complex. This result indicates that newly synthesized plasma membrane proteins and proteins destined for secretion are packaged into a single vesicle species in yeast.

Key words: transport, vesicle purification, clathrin, vacuolar targeting

The multicompartmental organization of eukaryotic cells requires that a protein be targeted specifically to its final destination. In several instances, this targeting clearly involves signals encoded by an amino acid sequence within the protein. For example, initial localization to the secretory pathway generally requires an N-terminal signal peptide that is cleaved during translocation of the protein into the endoplasmic reticulum (ER) [1]. Beyond this initial step, the mechanisms for sorting proteins to specific locations within the secretory pathway remain poorly understood. How are proteins retained within the ER and Golgi apparatus, exported to the plasma membrane or cell exterior, or

Received May 27, 1987.

diverted to the lysosome? Do all proteins that travel through the secretory pathway follow an identical route?

Study of the secretory process in the yeast *Saccharomyces cerevisiae* provides an opportunity for genetic manipulation that is currently unavailable in higher eukaryotic cells. A pathway of protein transport in yeast has been detected through the study of *sec* mutants that possess temperature-sensitive mutations that block secretion [2,3]. At the restrictive temperature, these mutants accumulate ER, Golgi, or secretory vesicles filled with secretory proteins. Double-mutant analysis has allowed ordering of secretory events and has revealed that secretion in yeast occurs by a process that is homologous to that in higher eukaryotes; proteins enter the ER, pass through the Golgi apparatus, and are packaged into secretory vesicles, which migrate to and fuse with the plasma membrane [4]. In addition, yeast contains a vacuolar compartment similar to the lysosome in its content of hydrolytic glycoprotein enzymes [5].

Using yeast as a model system, we performed a study to determine if various classes of protein follow an identical route through the secretory pathway. If so, all transported proteins should pass through the same secretory organelles. Alternatively, if routes deviate, organelles would contain subcompartments or subpopulations occupied by different proteins. Such segregation, if maintained throughout the secretory pathway, would be reflected in distinct classes of secretory vesicles. For this reason, secretory vesicles were purified and examined for content of the plasma membrane ATPase, the secretory protein acid phosphatase, and the vacuolar protein carboxypeptidase Y (CPY).

SECRETORY VESICLES FROM YEAST

Secretory vesicles have been isolated from mammalian endocrine, exocrine, and nerve cells, that store large numbers of vesicles until secretion is triggered by an external stimulus [for example, see 6–8]. In contrast, yeast cells secrete proteins by a constitutive process with the steady-state population of vesicles being very small. For this reason, purification of secretory vesicles from wild-type yeast has been quite difficult.

At the restrictive temperature, *sec1* mutant cells accumulate large quantities of Golgi-derived vesicles that contain the secretory enzyme acid phosphatase [2,4]. Therefore, this strain is potentially a good starting source for secretory vesicle purification. Novick and Schekman showed that the vesicles accumulated in *sec1* whole cells fuse with the plasma membrane upon return to the permissive temperature [2–4]. Since preliminary experiments demonstrated that higher yields of secretory vesicles could be obtained from yeast spheroplasts rather than whole cells, reversible accumulation of vesicles in *sec1* spheroplasts was examined (Table I). Accumulation of acid phosphatase was temperature-dependent, and secretion was restored upon return to the permissive temperature. Secretion of accumulated enzyme required energy, since reversible export was blocked in the presence of sodium azide. Therefore the vesicles of *sec1* spheroplasts are functional.

Isolation and Characterization

Secretory vesicles were isolated from spheroplasts by a series of differential and Percoll gradient centrifugations. The resulting preparations were enriched for secretory vesicles at least five-fold over other major cellular membranes [9,10] and were approximately 70% pure, as judged by electron microscopy [11]. Protein content of the secretory

TABLE I. Reversibility of Secretion Block in *sec1* Spheroplasts*

	Internal acid phosphatase (% total)			
	24°C		37°C	
NaN_3	−	+	−	+
No chase	20	20	80	56
With chase at 24°C	25	19	27	63

**sec1* Spheroplasts were pulse-labeled with $^{35}SO_4^{2-}$ at the permissive (24°C) or restrictive temperature (37°C) for 90 minutes [9]. Cycloheximide (0.4 mM) was added to halt protein synthesis. Sodium azide (10 mM) was added to half of each culture. Spheroplasts not subjected to a chase period were collected by centrifugation and stored on ice for 3 hours. The cultures that were subjected to a chase period were incubated at 24°C for 3 hours. Spheroplasts were collected and acid phosphatase from the spheroplasts (internal acid phosphatase) and from the culture supernatants (secreted acid phosphastase) was immunoprecipitated and quantified by autoradiography [10].

vesicles was analyzed by SDS polyacrylamide gel electrophoresis of $^{35}SO_4^{2-}$ metabolically labeled vesicles. Five major protein species of relative mobilities corresponding to 200, 125, 51, 33, and 20 kD, copurified with the secretory vesicles (Fig. 1). Of these proteins, three were firmly embedded in the vesicle membrane, as judged by lack of solubilization in alkaline solution (Fig. 2). Other protein species that were evident in this analysis did not copurify with the secretory vesicles; they may be secretory vesicle-associated proteins that are also present in other cellular compartments or contaminants of the vesicle preparations [9].

A 200-kD vesicle protein that was not integral to the membrane was examined further because these characteristics are similar to those of clathrin heavy chain (CHC), a protein of approximately 190 kD that is a major constituent of yeast coated vesicles [14,15]. To determine if clathrin was present in secretory vesicles, CHC was followed throughout our preparations (Table II). Only very small quantities of clathrin were found in preparations of secretory vesicles, and no enrichment of CHC was detected during the purification. From this, we conclude that the isolated secretory vesicles do not contain coats composed of CHC; hence the identity of the 200-kD protein remains unknown. The absence of CHC is consistent with several possibilities: the secretory vesicles may be clathrin coated in vivo but lose their coats during isolation; the vesicles may only be coated during budding from the Golgi but not exist as long-lived stable intermediates; or the vesicles may never be clathrin-coated. Unfortunately, the electron density of ribosomes within the yeast cytosol precludes direct examination of yeast thin sections for the presence of clathrin coats on secretory vesicles. It is unlikely that our preparation procedure stripped clathrin from the vesicles since none of the conditions we used are known to disrupt coats obtained from other sources. Furthermore, coated vesicles isolated from wild-type yeast cells sediment quite differently than secretory vesicles. It is also of note that vesicles carrying constitutively secreted proteins in mammalian cells appear to lack clathrin coats [reviewed in 16].

Types of Proteins Present

Preparations of secretory vesicles were examined for the presence of three types of proteins: newly synthesized plasma membrane, secretory, and vacuolar enzymes.

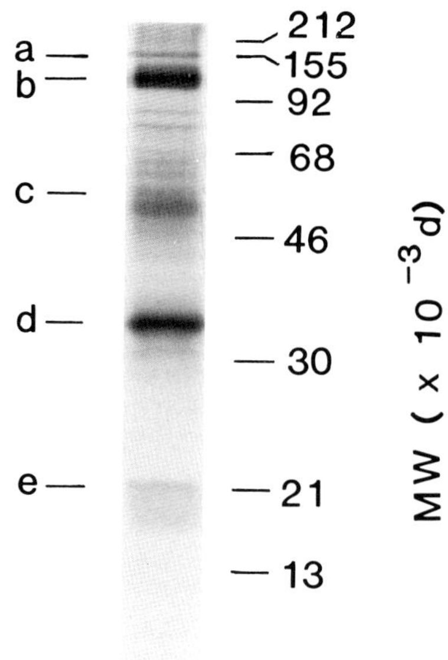

Fig. 1. Secretory vesicle-associated proteins. *sec1* Spheroplasts were labeled with $^{35}SO_4^{2-}$ at the restrictive temperature, and secretory vesicles were isolated from the Percoll gradient as described elsewhere [9]. This fraction was subjected to polyacrylamide gel electrophoresis (PAGE) [12] and visualized by autoradiography. The major polypeptides that copurified with the secretory vesicles are designated a–e.

Representative proteins from each class were quantified throughout the secretory vesicle isolation (Table III). The secretory vesicle purification procedure was devised using acid phosphatase as a marker. As expected, a large proportion of this marker was recovered in intact vesicles, and there was a substantial enrichment of acid phosphatase over total protein in the purest fraction.

The vacuolar protein CPY is synthesized as a proprotein that is cleaved to its mature form following transport through the early Golgi [17,18]. Greater than 95% of the CPY in *sec1* was found to be of mature form and occurred in fractions known to contain vacuolar proteins. These results confirm earlier findings by Stevens et al. [18] that processing of CPY proceeds in *sec1* cells and that this processed CPY may be present within vacuoles. In the previous studies, however, contamination of vacuolar preparations by secretory vesicles was not assessed. Our results indicate that CPY, a representative vacuolar protein, is sorted away from secretory proteins prior to the formation of the vesicles accumulated in *sec1*, and provide further evidence that CPY transport is not mediated by a secretion-recapture mechanism. This protein must travel to its final destination in organelles distinct from secretory vesicles. Such segregation of vacuolar proteins from secretory proteins is similar to the sorting of lysosomal enzymes in mammalian cells [19].

Any protein present in secretory vesicles should show purification properties similar to those of acid phosphatase. This was the case for ATPase (Table III). In every step of the isolation, the purification and yields of newly synthesized ATPase and acid phosphatase were identical [10]. This data suggested that these two types of proteins travel to the cell surface in the same population of vesicles.

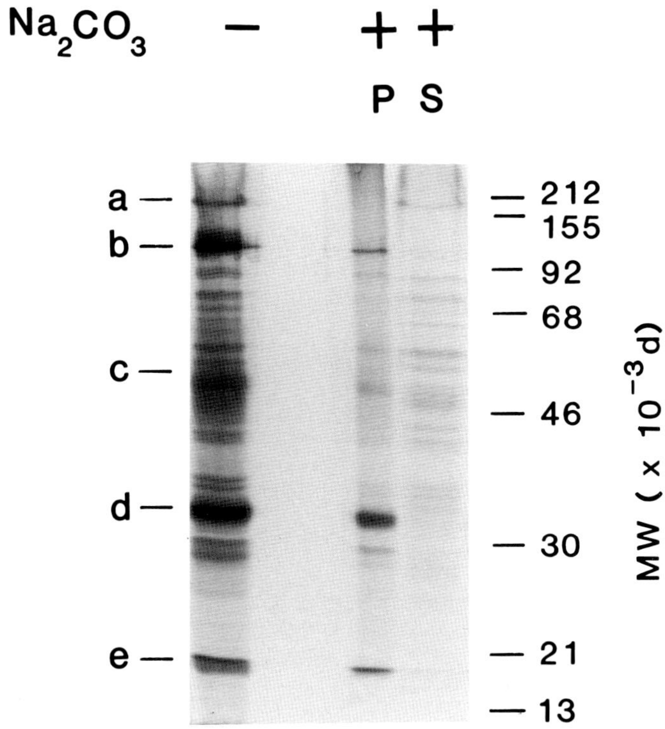

Fig. 2. Integral membrane proteins of secretory vesicles. Vesicles were purified from *sec1* spheroplasts as in Figure 1, washed with Na_2CO_3 to remove peripherally bound proteins [13], and collected by centrifugation. The pellet (P), containing integral membrane proteins, and supernatant (S) fractions were subjected to PAGE and visualized by autoradiography.

While the copurification of acid phosphatase and ATPase was consistent with their coexistence in the same vesicles, an alternative explanation was that they were present in distinct vesicles that copurified based on similar properties. To investigate this possibility further, we performed an experiment to determine if the vesicles containing newly synthesized ATPase also contained acid phosphatase. According to the structural model postulated for ATPase, a large portion of the polypeptide is exposed on the cytoplasmic

TABLE II. Distribution of Clathrin Heavy Chain in Secretory Vesicle Preparations*

Fraction	% Recovery	
	Acid phosphatase	CHC
Homogenate	100	100
Membrane pellet	63	15
Secretory vesicles	37	2

*Secretory vesicles from *sec1* spheroplasts grown in minimal media were isolated from the Percoll gradient as described elsewhere [10]. Acid phosphatase was used as a secretory vesicle marker. Clathrin heavy chain was quantified by immunoblot [10]. Total yield for each marker was at least 90% in all purification steps.

TABLE III. Types of Proteins in Secretory Vesicles*

	Relative specific activity (% recovery)		
Fraction	Acid phosphatase	ATPase	CPY
Homogenate	1 (100)	1 (100)	1 (100)
Secretory vesicles	31 (37)	25 (31)	1 (3)

**sec1* Spheroplasts were labeled with $^{35}SO_4^{2-}$ at the restrictive temperature, and secretory vesicles were isolated from the Percoll gradient as described elsewhere [9]. All marker proteins were quantified by immunoprecipitation and autoradiography. Total yield for each marker was at least 85% in all purification steps.

surface of the plasma membrane [20]. Since membrane topology is maintained throughout export, the vesicles in which ATPase is accumulated in *sec1* should have the same portion of ATPase exposed on their external surface. This portion of the ATPase could be recognized by appropriate antibody and could be complexed to *Staphylococcus aureus* cells via this antibody. Under nondenaturing conditions, recognition by the antibody would complex the entire ATPase-containing vesicle. If ATPase resides in secretory vesicles, then acid phosphatase would also be present in the complex. Results from such an experiment are shown in Table IV. Both ATPase and acid phosphatase were recovered quantitatively in the *S. aureus* complex. When nonimmune serum was used, neither protein was present in the complex. Thus the recognition of vesicles was mediated specifically via the ATPase antibody. Hence a plasma membrane protein and a secretory protein reach the cell surface in the same secretory vesicles. This implies that the two types of protein follow an identical route through the entire secretory pathway.

In mammalian cells infected with vesicular stomatitis virus, the G protein, a viral integral-membrane glycoprotein, has been detected in vesicles (assumed to be secretory vesicles) that also contained the secretory proteins transferrin and albumin [21]. The existence of a small number of such vesicles in thin sections of cells does not rule out the possibility that the bulk of the secretory and plasma membrane proteins are sorted, and carried to the cell surface in vesicles distinct from those observed. In contrast, our studies employed a yeast mutant in which delivery of plasma membrane and secretory proteins was blocked completely [2,10], thus allowing quantitative localization of pro-

TABLE IV. Coimmunoprecipitation of ATPase and Acid Phosphatase*

	% in complex	
Serum	ATPase	Acid phosphatase
Anti-ATPase	98	94
Nonimmune	4	8

**sec1* Spheroplasts were labeled with $^{35}SO_4^{2-}$ at the restrictive temperature, and secretory vesicles were isolated from the Percoll gradient as described elsewhere [9]. Vesicles were incubated with *S. aureus* that had been pretreated with the appropriate serum [10]. The *S. aureus* was collected by centrifugation. ATPase and acid phosphatase in the pellet (complexed) and supernatant were quantified by immunoprecipitation and autoradiography.

teins. Our results provide evidence that constitutively secreted integral membrane and secretory proteins are packaged into a single uniform population of vesicles.

PROSPECTS

The availability of purified secretory vesicles from yeast provides the opportunity to address numerous issues pertaining to secretion. A molecular description of the last step in the secretory pathway will require reconstitution of the fusion of secretory vesicles with vesicles derived from the plasma membrane. The secretory vesicle isolation procedure is of obvious significance in providing one of the reagents for this system. However, applicability of the fractionation procedure extends beyond this. A protein involved in this final secretory step would be expected to fit into one of three classes: cytosolic, secretory vesicle, or plasma membrane component. Our secretory vesicle enrichment procedure effectively separates these classes of protein. Hence the cellular location of proteins already identified as being involved in transport from the Golgi apparatus to the plasma membrane (i.e., *SEC* gene products required in vesicle discharge) can be determined. Localization of the various *SEC* gene products may expedite the development of an in vitro system by allowing the combination of fractions from various mutant cells so as to include wild-type fractions of secretory vesicles, plasma membrane, and cytosol. Furthermore, knowledge of the cellular location and sequence of a SEC protein involved in vesicle discharge may allow its function to be predicted. Models based on this information may be tested by methods that do not require a reconstituted fusion event. For example, a *SEC* gene product that is integral and spans the secretory vesicle membrane in several loops may mediate pH or ionic changes within the vesicle. A component of the plasma membrane may behave as a vesicle receptor or fusogen. Components present in the cytosol that become associated with secretory vesicles under specific conditions may be homologs of mammalian chromobindins [22] or may be active in forming coat structures not composed of clathrin [23].

The availability of purified vesicles may allow isolation of additional *sec* mutants. In yeast, the function of a polypeptide may be assessed through purification, molecular cloning of the corresponding gene, and the creation of chromosomal mutations that allows the analysis of resulting defects. In this way, a given secretory vesicle-associated protein may be shown to play a crucial role in secretion.

A further application of the vesicle isolation procedure is its usefulness in studying processing events within the secretory pathway. Numerous secretory proteins are synthesized as precursors that are proteolytically processed during export. In mammalian cells, proinsulin processing has been shown to occur within secretory granules [24]. To date, the cellular location of homologous processing in yeast is unknown. Examination of the secretory vesicles, both biochemically and by immunoelectron microscopy for the presence of processing enzymes should reveal if these events occur within vesicles.

ACKNOWLEDGMENTS

This work was supported by grants NSF DCB8402552 and NIH GM36881. We thank G. Schatz for the gift of antiserum to acid phosphatase. We are grateful to N. Nelson for helpful suggestions and antiserum to ATPase. And we extend special thanks to G. Payne for providing antiserum to CHC and for his support throughout the project.

REFERENCES

1. Kreil G: Ann Rev Biochem 50:317, 1981.
2. Novick P, Schekman R: Proc Natl Acad Sci USA 76:1858, 1979.
3. Novick P, Field C, Schekman R: Cell 21:205, 1980.
4. Novick P, Ferro S, Schekman R: Cell 25:461, 1981.
5. Weimken A, Schellenberg M, Urech K: Arch Microbiol 123:23, 1979.
6. Gumbiner B, Kelly RB: Proc Natl Acad Sci USA 78:318, 1981.
7. Carlson SS, Wagner JA, Kelly RB: Biochemistry 17:1188, 1978.
8. Helmy S, Porter-Jordan K, Dawidowicz EA, Pilch P, Schwartz AL, Fine RE: Cell 44:497, 1986.
9. Holcomb CL, Etcheverry T, Schekman R: Anal Biochem 166:328, 1987.
10. Holcomb CL, Hansen W, Etcheverry T, Tschopp J, Schekman R: J Cell Biol (in press), 1987.
11. Holcomb CL, Brada D: Unpublished observation.
12. Laemmli UK: Nature 227:680, 1970.
13. Fujiki Y, Hubbard AL, Fowler S, Lazarow PB: J Cell Biol 93:97, 1982.
14. Mueller SC, Branton D: J Cell Biol 98:341, 1984.
15. Payne GS, Schekman R: Science 230:1009, 1985.
16. Griffiths G, Simons K: Science 234:438, 1986.
17. Hasilik A, Tanner W: Eur J Biochem 85:599, 1978.
18. Stevens T, Esmon B, Schekman R: Cell 30:439, 1982.
19. Sly WS, Fischer HD: J Cell Biochem 18:67, 1982.
20. Serrano R, Kielland-Brandt MC, Fink GR: Nature 319:689, 1986.
21. Strous GJAM, Willemsen R, van Kerkhof P, Slot JW, Geuze HJ, Lodish HF: J Cell Biol 97:1815, 1983.
22. Creutz CE, Zaks WJ, Hamman HC, Crane S, Martin WH, Gould KL, Oddie KM, Parson SJ: J Biol Chem 262:1860, 1987.
23. Orci L, Glick BS, Rothman JE: Cell 46:171, 1986.
24. Orci L, Ravazzola M, Amherdt M, Madsen O, Vassalli JD, Perrelet A: Cell 42:671, 1985.

Molecular Biology of Intracellular Protein
Sorting and Organelle Assembly 161–171 (1988)

Mutational Analysis of the Yeast F_1-ATPase β-Subunit Mitochondrial Protein Import Signal: Evidence of Functional Redundancy

David M. Bedwell and Scott D. Emr

Division of Biology, California Institute of Technology, Pasadena, California 91125

Many proteins destined for mitochondria are initially synthesized with a transient NH_2-terminal presequence that directs import of the protein into mitochondria. In an attempt to define the functional characteristics of import signals, we have utilized site-directed mutagenesis to introduce a series of mutations into the NH_2-terminal portion of an imported yeast mitochondrial protein, the F_1-ATPase β-subunit (encoded by the *ATP2* gene). Plasmids encoding the mutant proteins were transformed into a Δ*atp2* strain, and the import properties of the mutant F_1 β-subunit proteins were examined. We found that a combination of two nonoverlapping deletions within the presequence (Δ5–12 and Δ16–19) that removed all four basic residues (along with 12 of the 19 amino acids) in the presequence did not significantly block import into mitochondria, suggesting that additional targeting information is present within the mature protein. To test this model, the entire presequence (except the NH_2-terminal methionine) was deleted. We found that this "mature" F_1 β-subunit protein was still competent for import into mitochondria in vivo. Import was abolished only when the two presequence deletions were combined with a third deletion (Δ28–34) that removed two more basic residues within the mature F_1 β-subunit protein. Examination of various constructions containing all possible pairwise combinations of the three deletions showed that the combination of any two deletions did not disrupt import. These results indicate that the F_1 β-protein contains redundant import information and separate regions near the NH_2-terminus of the F_1 β-subunit precursor (including one in the mature protein) are capable of supplying a function necessary for the import of β into mitochondria.

Key words: site-directed mutagenesis, redundant targeting information, targeting mutants, *Saccharomyces cerevisiae*

The transport of proteins into and across biological membranes is essential for living cells to maintain their high level of compartmentalization. One example of the importance of this process is the import of proteins into mitochondria. More than 90% of mitochondrial proteins are encoded by nuclear DNA and synthesized in the cytoplasm

Received June 15, 1987.

before being imported to one of four distinct mitochondrial compartments: the outer membrane, the intermembrane space, the inner membrane, or the matrix [1]. Both the sheer volume of proteins utilizing this transit process and the apparent high specificity with which these proteins are directed to their correct submitochondrial location indicate that an extremely efficient and accurate mechanism is responsible for these events.

A schematic outline of the import of a typical mitochondrial protein, the F_1-ATPase β-subunit, is shown in Figure 1A. This protein is synthesized on cytoplasmic ribosomes in a precursor form that contains a transient NH_2-terminal presequence [2]. The precursor binds to the mitochondrial surface, presumably through an interaction with proteinaceous receptors [3] that are presently still uncharacterized. The bound protein is then transferred across the outer and inner mitochondrial membranes. Once within the mitochondrial matrix, the presequence is proteolytically removed, and the mature protein is assembled into a functional proton-translocating ATPase complex. Recent work has resolved the energy-dependent portion of the import process into two distinct steps [4–6], as seen in Figure 1B. In the first step, the presequence of the bound precursor is inserted into the inner membrane by a process strictly dependent upon a potential across the inner membrane. Once this insertion is achieved, the membrane potential is no longer required, and the remainder of the protein is translocated across the membrane in an ATP-dependent process. Based upon this model, the primary (or perhaps only) role of the transient presequence is to direct the initial membrane potential-dependent insertion event.

Comparison of a large number of mitochondrial presequences has found no consensus primary amino acid sequence homology, although most are rich in basic and hydroxylated residues and free of acidic residues [7]. It has also been proposed that presequences can form amphiphilic helices containing high hydrophobic moments, which are thought to mediate their insertion into the mitochondrial inner membrane [8], possibly by an electrophoretic process in which the positively charged presequence is drawn into the membrane by the electronegative-inside polarity of the inner membrane potential. However, while artificial presequences designed to form an amphiphilic helical structure can direct import, other sequences expressly designed to distribute the basic residues around the helical presequence also promote mitochondrial import [9]. Therefore the functionally important features of these import signals are still not clearly defined.

The import properties of the F_1-ATPase β-subunit have been characterized both in vivo [10,11] and in vitro [2,12]. This protein is initially synthesized with a 19-amino acid presequence that is removed upon its arrival in the mitochondrial matrix [13]. Previous gene fusion and deletion experiments have shown that the mitochondrial targeting information of the F_1 β-subunit precursor, like other mitochondrial proteins, is located near its NH_2-terminus [11,13,14]. This knowledge makes the F_1 β-subunit an ideal candidate for a more detailed examination of the functional components of a mitochondrial import signal.

In this report we describe the use of oligonucleotide-directed mutagenesis to alter various regions in or near the F_1-ATPase β-subunit presequence. Using this approach, we have found evidence of functionally redundant mitochondrial import information at the NH_2-terminus of this protein.

F_1-ATPase β-SUBUNIT CONTAINS REDUNDANT IMPORT INFORMATION

The set of F_1 β-subunit mutants analyzed in this study are shown in Figure 2. Our initial strategy was to alter or remove various portions of the presequence that contain

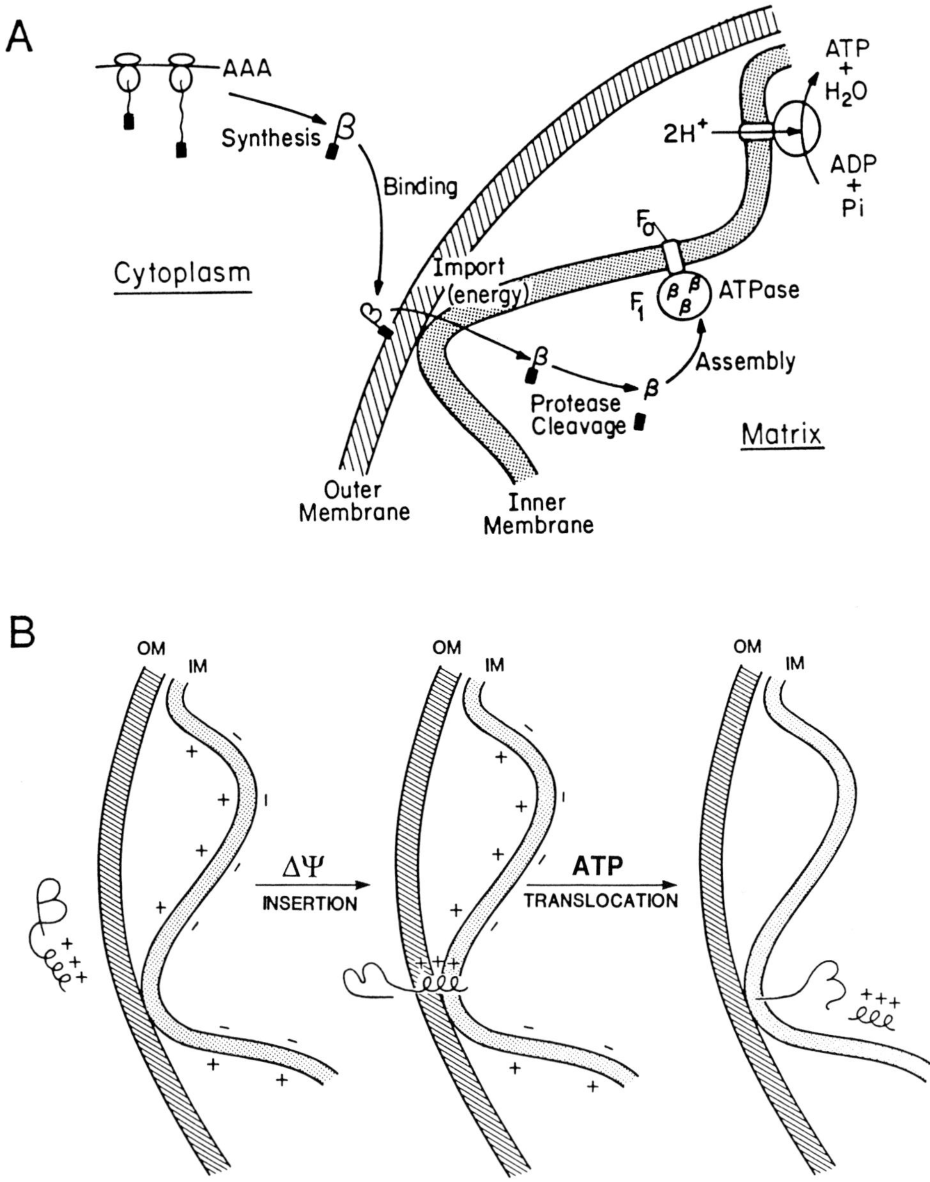

Fig. 1. Mitochondrial import. **A)** Schematic representation of general steps in mitochondrial protein import. **B)** Energy-dependent steps of mitochondrial protein import.

charged amino acid residues. Each of these mutations was introduced into a copy of the F_1 β-gene (*ATP2*) on a low copy number ARS CEN plasmid vector and was then transformed into the yeast strain SEY6215, which is deleted for the chromosomal copy of the F_1 β-subunit gene (Δ*atp2*). An intact mitochondrial proton-translocating ATPase complex is essential for growth on nonfermentable carbon sources such as glycerol or lactate. Therefore an indication of F_1 β-subunit import into mitochondria can be obtained by expressing each of the mutant F_1 β-subunit proteins in the Δ*atp2* strain and testing

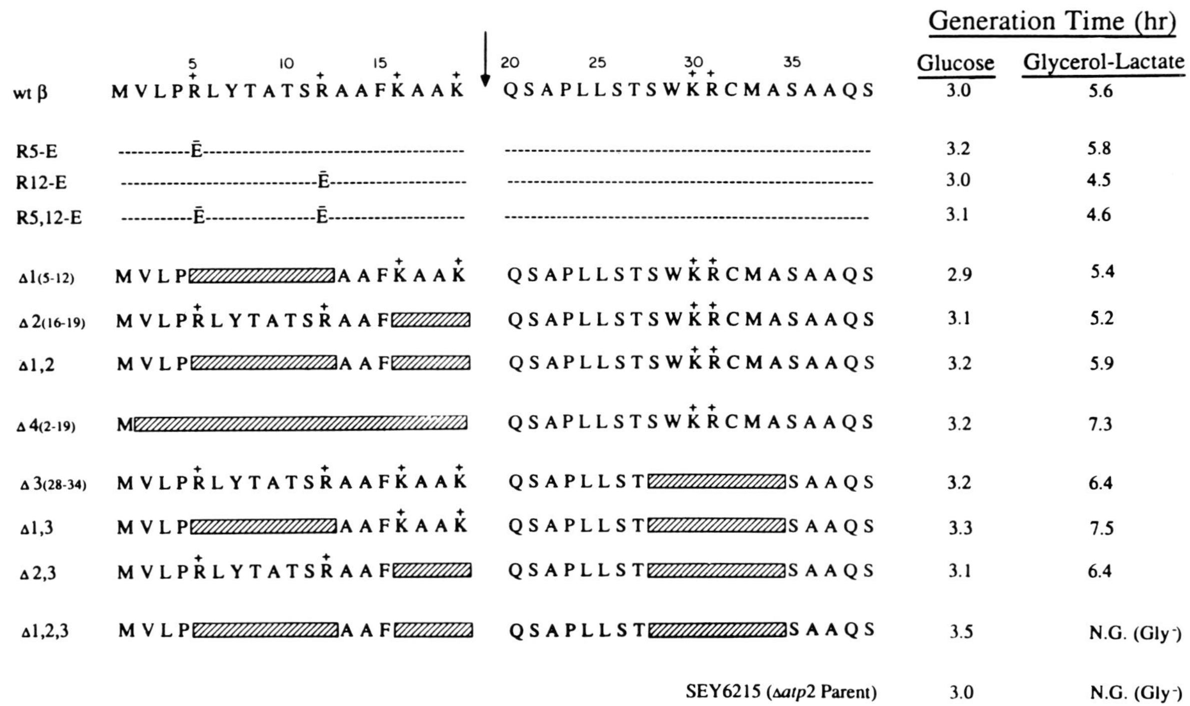

Fig. 2. Mutant F_1 β-subunit proteins. The NH_2-terminal residues of the wild-type F_1 β-subunit precursor [23] and mutant derivatives as predicted from the nucleotide sequence. For missense mutants (R5-E, R12-E, and R5,12-E), dashed lines represent wild-type sequences. Crosshatched boxes represent deletions. Basic residues are indicated with a plus sign; acidic residues, with a minus sign. The arrow above the sequence indicates the proteolytic-processing site [15]. After transforming each of the centromere vectors containing the DNA coding sequences of these F_1 β-subunit constructions into the Δ*atp2* strain SEY6215, growth rates were determined in both YNB medium supplemented with 2% glucose (a fermentable carbon source) and YP medium supplemented with 3% glycerol and 3% DL-lactate (nonfermentable carbon sources). N.G., no growth; Gly^-, $Glycerol^-$ phenotype.

for growth on media containing only glycerol and lactate as carbon sources. In fact, this growth assay can also give an indication of the relative efficiency of these mutant mitochondrial import signals, since it has been shown that the inability to efficiently target the F_1 β-subunit to mitochondria can lead to a severe growth rate reduction on a nonfermentable carbon source [15].

Upon examination of the growth characteristics of the strains expressing the mutant F_1 β-subunit proteins, it was found that neither single- or double-point mutations substituting glutamic acid residues for arginine residues (R5-E, R12-E, and R5,12-E) reduced growth rates on glycerol/lactate media below wild-type levels (see Fig. 1). In addition, deletions removing pairwise combinations of basic residues (Δ1 and Δ2), or all four basic residues of the sequence (Δ1,2), also caused no changes in growth rates relative to the wild-type control. These results were surprising, based upon the strong bias toward basic residues (and against acidic residues) found in mitochondrial targeting signals in general. Because of these results, we questioned whether the F_1 β-subunit presequence was essential for the import of this protein into mitochondria. To test this idea, an additional mutant (Δ4) was constructed that deleted the entire presequence except for the initiating methionine residue. It was found that the Δ4 mutant construction was also able to restore growth of the Δ*atp2* strain on glycerol/lactate media, indicating that additional targeting information must be present within the mature F_1 β-subunit protein. We noted that two additional basic residues are present near the NH_2-terminus of the mature protein at residues 11 and 12 (residues 30 and 31 of precursor), so this region was also deleted. We found that growth on nonfermentable carbon sources indeed was abolished when a deletion removing residues 9–15 of the mature protein (Δ3) was combined with deletions removing most of the presequence (Δ1,2,3). This result confirmed that additional targeting information resides near the NH_2-terminus of the mature F_1 β-subunit protein. A systematic analysis of pairwise combinations of deletions Δ1, Δ2, and Δ3 (Δ1,2; Δ1,3; and Δ2,3) revealed that each of these double mutants retained the ability to grow on nonfermentable carbon sources, indicating that various portions of the NH_2-terminus can direct mitochondrial import. Since the slowest growth rate of any single or double mutant on glycerol/lactate media was no more than 35% slower than the F_1 β-subunit wild-type control, F_1 β-subunit import apparently is not reduced much below the level required for efficient growth on a nonfermentable carbon source. These results indicate that the F_1 β-subunit precursor contains redundant mitochondrial protein import information, and these redundant regions can direct functionally equivalent steps in the mitochondrial import process.

LOCALIZATION OF MUTANT F_1 β-SUBUNIT PRECURSORS TO MITOCHONDRIA

In order to quantitate the amount of F_1 β-subunit imported into mitochondria, strains containing various mutant F_1 β-subunit constructions were grown to mid-log phase and harvested, and mitochondria were isolated [16]. Immunoblot analysis of these preparations was then done to determine the amount of F_1 β-subunit protein within mitochondria. It was found that all but the Δ1,2,3 mutant had a normal amount of F_1 β-subunit protein within the mitochondrial fraction (Fig. 3). When these fractions were assayed for mitochondrial ATPase activity, most had near wild-type levels of activity, confirming that the protein was localized to the mitochondria and assembled into a functional ATPase complex. Again, only the Δ1,2,3 mutant had little or no ATPase

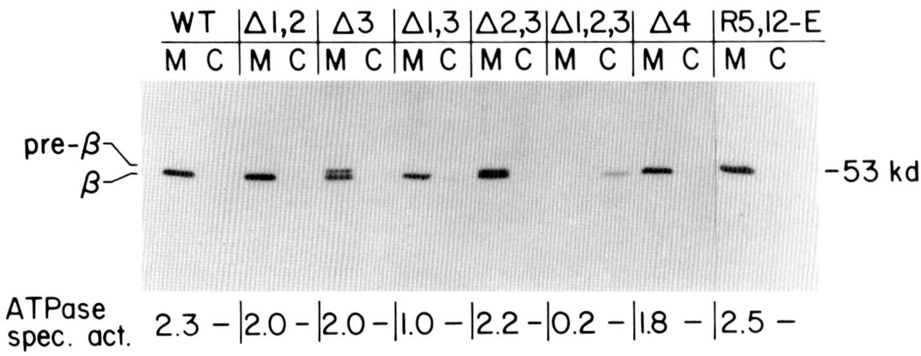

Fig. 3. Subcellular localization of mutant F_1 β-subunit proteins. Cultures of SEY6215 expressing wild-type or mutant F_1 β-subunit proteins were grown in YNB medium containing 2% galactose as carbon source. Cells were enzymatically treated to form spheroplasts and fractionated into mitochondrial (M) and cytoplasmic (C) fractions (the cytoplasmic fraction is equivalent to the postmitochondrial supernatant); cell equivalent amounts of these fractions were electrophoresed on SDS-polyacrylamide gels and immunoblotted. Mitochondrial ATPase specific activities are expressed as nmoles/min/mg protein and are not corrected for background (the mitochondrial ATPase specific activity of SEY6215 was ~ 0.1 nmole/min/mg protein).

activity, indicating that the F_1 β Δ1,2,3 mutant protein is not efficiently imported into mitochondria. Since the other mutant F_1 β-subunit proteins are present in normal amounts, the mutations apparently do not disrupt F_1 β-subunit targeting enough to alter their steady-state accumulation in mitochondria.

We next wanted to examine the rate of mitochondrial import of the mutant F_1 β-subunit precursors. Since precursor maturation is dependent upon arrival in the mitochondrial matrix, the rate of precursor maturation is a reasonable measure of the rate of import [10,17]. Therefore pulse/chase experiments were done to determine the relative rate of precursor maturation (Fig. 4). It was found that both the wild-type control and most of the mutants examined showed little or no F_1 β-subunit precursor after a 5-minute pulse-labeling period (without chase), indicating rapid precursor maturation. Only two mutants (Δ1,2 and R5,12-E) showed significant amounts of F_1 β-subunit precursor still present after a 5-minute pulse with no chase. However, after a 5-minute chase period, essentially all of the precursor had matured, indicating that mitochondrial import of these mutant F_1 β-subunit proteins is only three- to five-fold slower than the wild-type F_1 β-subunit precursor rate of import [10]. Unfortunately, precursor maturation was not a useful measure of import for any of the mutant proteins containing the Δ3 mutation [13; our unpublished results] or the Δ4 mutation, which lacks a cleavable presequence. Therefore estimates of the relative rate of import of these mutants remain to be determined.

EXAGGERATION OF IMPORT DEFECTS IN ρ^- YEAST

It previously has been shown that a potential across the mitochondrial inner membrane is required for import of mitochondrial proteins [18]. This potential is thought to be required for the insertion of the amphiphilic presequence into the membrane, thus initiating the movement of the protein into the mitochondrion. Based upon the essential

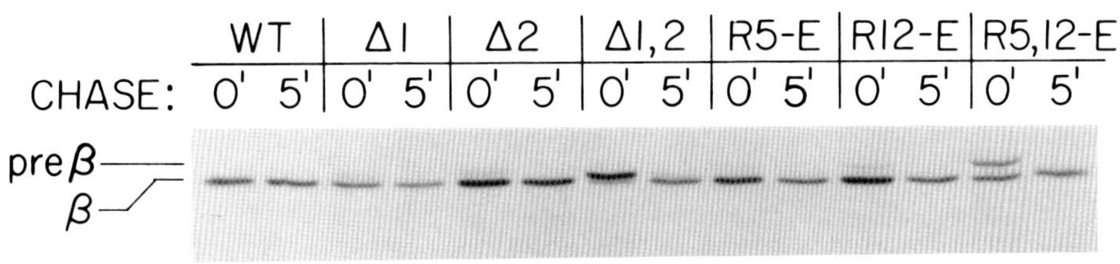

Fig. 4. Processing kinetics of mutant F_1 β-subunit proteins. SEY6215 cells expressing wild-type or mutant F_1 β-subunit proteins were labelled 5 minutes with $Na_2{}^{35}SO_4$, followed by a chase in the presence of 10 mM $(NH_4)_2SO_4$ and 100 μg/ml cycloheximide. The chase was terminated at the indicated times by the addition of TCA, followed by immunoprecipitation with F_1 β-subunit specific antisera.

nature of this membrane potential and its postulated direct involvement with the function of mitochondrial import signals, we reasoned that by reducing the membrane potential we might be able to exaggerate the import defect of some or all of the F_1 β-subunit mutants.

Under aerobic growth conditions, the mitochondrial membrane potential is generated by the electron transport chain, which pumps protons out of mitochondria. Under growth conditions where the electron transport chain is inactive, the membrane potential is due primarily to ATP hydrolysis by the mitochondrial ATPase complex, a process that also results in protons being pumped out of organelle. Both of these proton pumping mechanisms are simultaneously abolished in respiration-deficient rho$^-$ (ρ^-) yeast [19]. Because of the loss of these processes, ρ^- yeast strains apparently have a diminished mitochondrial membrane potential. We proceeded to utilize this "suboptimal" mitochondrial import condition to further characterize the import of the mutant F_1 β-subunit proteins.

To examine the import of our mutant F_1 β-subunit constructions under these conditions, a Δ*atp2*, ρ^- strain was constructed, and plasmids encoding the mutant F_1 β-subunit proteins were introduced. When growth rates of these strains were examined, it was found that they could be grouped into two classes (see Table I). The first class had growth rates similar to the ρ^-, βWT control strain, while the second class had growth rates three-fold slower (similar to the ρ^-, Δ*atp2* control). This result suggested that the class 2 mutants may be unable to efficiently import the mutant F_1 β-protein due to an exaggeration of their import defect in the ρ^- background.

We next examined the rate of precursor maturation of the class 1 ρ^- mutants (Fig. 5). It was found that the rate of precursor maturation for these mutants is clearly delayed in the ρ^- background, relative both to the ρ^-, βWT control and to the same mutants in a ρ^+ strain (compare with Fig. 4). This result indicates that import of these class 1 mutants is slower in ρ^- cells. In order to test the model that the class 2 mutants are not efficiently targeted in the ρ^- background, mitochondria were isolated from three ρ^- strains, two class 1 strains (βWT and βΔ1), and a class 2 strain (βΔ1,2). Immunoblot analysis revealed that F_1 β-subunit could be detected in the mitochondrial fraction of both class 1 strains, but not in the class 2 strain (data not shown). From these results we conclude that the class 2 mutants encode F_1 β-subunit precursor proteins that are not competent for efficient import into ρ^- yeast, even though they are capable of directing import in ρ^+ yeast.

TABLE I. Growth Rates of *Rho*$^-$ Strains*

ATP2 allele	Doubling time (hrs)
Class 1	
βWT	2.2
βR5-E	2.1
βR12-E	2.1
βR5,12-E	3.2
βΔ1	2.2
βΔ2	2.2
βΔ3	2.5
Class 2	
Δ*atp2*	7.3
βΔ1,2	7.0
βΔ1,3	6.5
βΔ2,3	6.3
βΔ1,2,3	6.3
βΔ4	6.3

*SEY6215 ρ^- cells expressing the indicated F_1 β-subunit mutant proteins were inoculated from plates that maintained a positive selection for the plasmid-encoded URA3 marker and grown in YP medium containing 8% glucose at 30°C. Growth rates were determined by measuring the increase in cell density as determined by OD_{600} measurements.

IN VITRO IMPORT OF MUTANT F_1 β PRECURSORS

It previously was reported that the mature F_1 β-subunit protein is not competent for import into mitochondria in vitro [2]. However, we have presented data indicating that our Δ4 mutant (deleted for the entire presequence except the NH_2-terminal methionine) is efficiently accumulated in mitochondria in vivo. To address this discrepancy, we have also examined the import of this mutant precursor into mitochondria in vitro. Precursor protein was synthesized in an SP6-directed transcription/rabbit reticulocyte

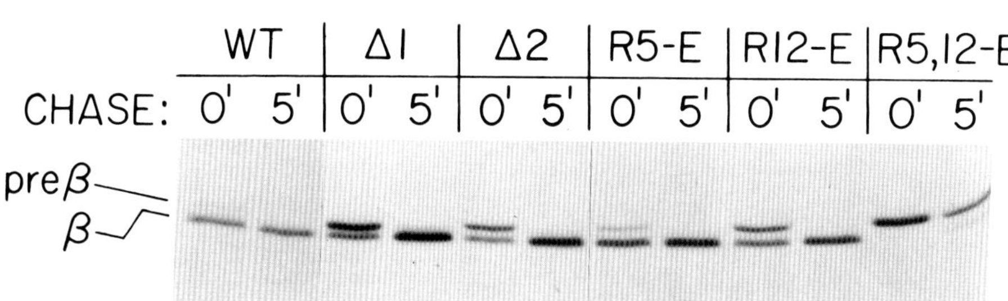

Fig. 5. Processing kinetics of mutant F_1 β-subunit proteins in ρ^- yeast. SEY6215 ρ^- cells expressing wild-type or mutant F_1 β-subunit precursors were grown in Wickerham's minimal medium containing 8% glucose, labeled 5 minutes with $Na_2{}^{35}SO_4$, and chased in the presence of 10 mM $(NH_4)_2SO_4$ and 100 μg/ml cycloheximide for the indicated time period. The chase was terminated by the addition of TCA, followed by immunoprecipitations using F_1 β-subunit specific antisera.

translation system and incubated with isolated mitochondria. We found that the Δ4 mutant precursor was not competent for import into mitochondria in vitro. In fact, in other experiments we found that mutants defined as class 1 mutants in the ρ^- experiments were capable of in vitro mitochondrial import, while class 2 mutants were all unable to import in vitro (unpublished results). This suggests that the conditions of in vitro import may generally represent "suboptimal" conditions, as compared with the in vivo conditions that facilitate import of the Δ4 mutant protein. Therefore the in vitro import data most closely approximates mitochondrial import in the ρ^- background.

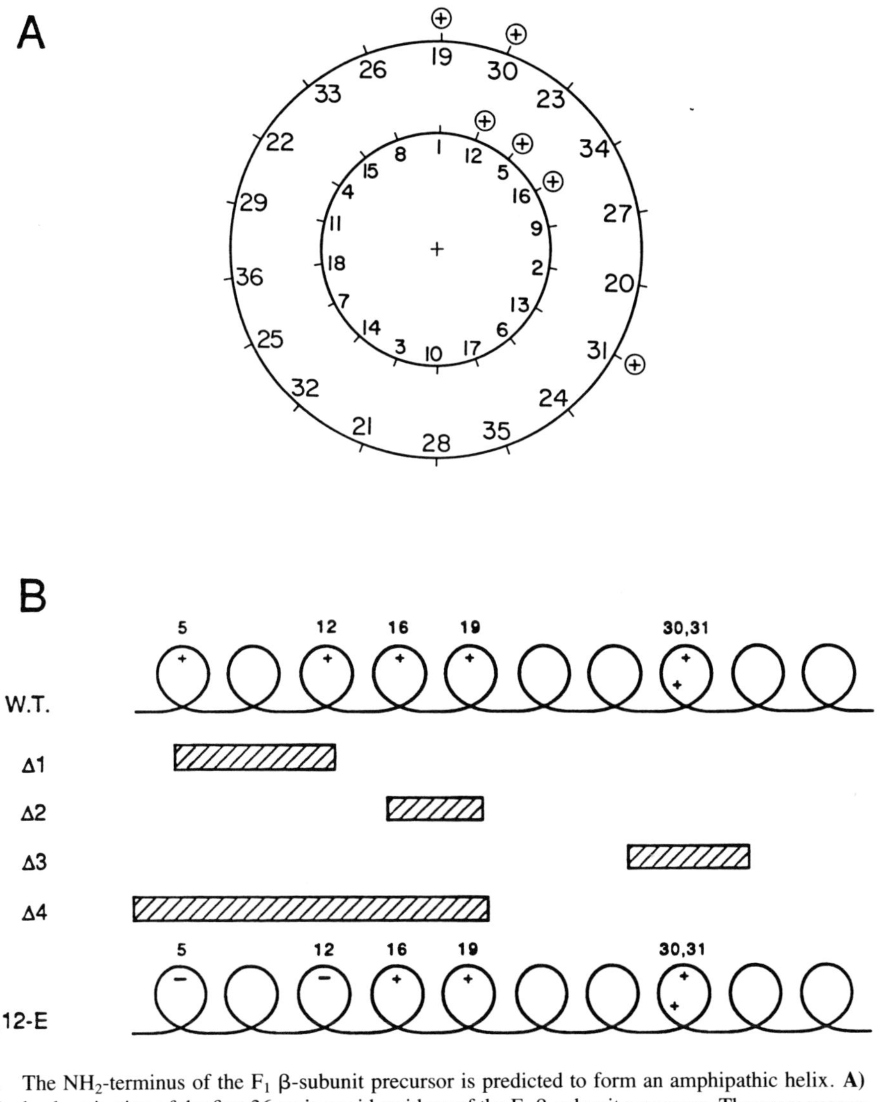

Fig. 6. The NH_2-terminus of the F_1 β-subunit precursor is predicted to form an amphipathic helix. **A)** Helical wheel projection of the first 36 amino acid residues of the F_1 β-subunit precursor. The presequence consists of the first 19 residues. **B)** Schematic representation of the location of F_1 β-subunit mutations in relation to the basic residues found within the NH_2-terminal 36 amino acid α-helical region involved in mitochondrial import.

CONCLUSIONS

A method of Fourier analysis was recently used by Vassarotti et al. [15] to predict that the NH_2-terminal 38 amino acids of the F_1 β-subunit precursor can form an amphipathic helix upon interaction with a membrane surface. This structure has previously been identified as a common element of most mitochondrial import signals [8]. Our results indicate that a combination of three small deletions within this region (removing a total of 19 amino acid residues) is necessary to block F_1 β-subunit import, while all possible pairwise combinations of these deletions do not significantly effect import in ρ^+ yeast. This indicates that various segments of this targeting signal can independently participate in directing functionally equivalent steps in the import process (see Fig. 6). Furthermore, we found that some mutant constructs that are sufficient for import in ρ^+ yeast are not able to direct import in ρ^- yeast. This conditional import phenotype suggests that the redundant regions of the import signal may act in an additive (or cooperative) manner to facilitate import of the F_1 β-subunit protein into mitochondria under suboptimal conditions.

We have observed a strong correlation between the net positive charge of the mutant F_1 β-subunit import signals and their ability to direct import into mitochondria. It is tempting to speculate that this corresponds to an important role for the multiple basic residues (and absence of acidic residues) generally found in mitochondrial presequences. Mitochondria are the only intracellular organelle having a membrane potential-oriented electronegative inside. The positively charged import signals on mitochondrial proteins may be able to detect this uniquely oriented potential and thereby use this distinguishing feature of the organelle to identify it as the correct delivery target. It has been shown that this electrochemical potential contributes to the membrane insertion of mitochondrial protein import signals [18], possibly by drawing the signal across the membrane by an electrophoretic process. Consistent with this model, the reduced net positive charge associated with some mutant F_1 β-subunit import signals may only be sufficient to allow this process to occur in ρ^+, but not ρ^- yeast.

The mitochondrial targeting signals of three other proteins have also been analyzed using various mutagenesis procedures. Studies of both the ornithine transcarbamylase precursor [20] and the cytochrome oxidase subunit IV precursor [21] concluded that only one portion of the presequence of each of these proteins could direct mitochondrial targeting. However, a study of the alcohol dehydrogenase III precursor [22] found that redundant import information may also be present within that protein. Therefore some mitochondrial precursors appear to contain redundant information, while other do not. Further studies are needed to determine why these differences occur.

REFERENCES

1. Douglas MG, McCammon MT, Vassarotti A: Microbiol Rev 50:166–178, 1986.
2. Maccecchini ML, Rudin Y, Blobel G, Schatz G: Proc Natl Acad Sci USA 76:343–347, 1979.
3. Zwizinski C, Schleyer M, Neupert W: J Biol Chem 259:7850–7856, 1984.
4. Pfanner N, Neupert W: FEBS Lett 209:152–156, 1986.
5. Eilers M, Oppliger W, Schatz G: EMBO J 6:1073–1077, 1987.
6. Chen WJ, Douglas MG: Cell 49:651–658, 1987.
7. Reid GA: Curr Top Membr Transport 24:295–336, 1985.
8. von Heijne G: EMBO J 5:1335–1342, 1986.

9. Allison DS, Schatz G: Proc Natl Acad Sci USA 83:9011–9015, 1986.
10. Reid GA, Schatz G: J Biol Chem 257:13062–13067, 1982.
11. Emr SD, Vassarotti A, Garrett J, Geller BL, Takeda M, Douglas MG: J Cell Biol 102:523–533, 1986.
12. Gasser S, Daum G, Schatz G: J Biol Chem 257:13034–13041, 1982.
13. Vassarotti A, Chen WJ, Smagula C, Douglas MG: J Biol Chem 262:411–418, 1987.
14. Douglas MG, Geller BL, Emr SD: Proc Natl Acad Sci USA 81:3983–3987, 1984.
15. Vassarotti A, Stroud R, Douglas M: EMBO J 6:705–711, 1987.
16. Daum G, Böhni PC, Schatz G: J Biol Chem 257:13028–13033, 1982.
17. Schleyer M, Neupert W: Cell 43:339–350, 1985.
18. Pfanner N, Neupert W: EMBO J 4:2819–2825, 1985.
19. Tzagoloff A: "Mitochondria." New York: Plenum Press, 1982.
20. Horwich AL, Kalousek F, Fenton WA, Pollock RA, Rosenberg LE: Cell 44:451–459, 1986.
21. Hurt E, Allison DS, Müller U, Schatz G: J Biol Chem 262:1420–1424, 1987.
22. Pilgrim D, Young ET: Mol Cell Biol 7:294–304, 1987.
23. Takeda M, Vassarotti A, Douglas MG: J Biol Chem 260:15458–15465, 1985.

Molecular Biology of Intracellular Protein Sorting and Organelle Assembly 173–186 (1988)

Vacuolar Protein Targeting and Organelle Biogenesis in Yeast

Daniel J. Klionsky, Lois M. Banta, Jane S. Robinson, and Scott D. Emr

Division of Biology, California Institute of Technology, Pasadena, California 91125

The yeast vacuole contains a number of hydrolytic glycoprotein enzymes, including proteinase A (PrA) and carboxypeptidase Y (CPY). Delivery of these proteins to the vacuole is mediated by a portion of the secretory pathway. Extensive work has been done on CPY to determine how the protein is targeted to the vacuole. We have begun additional studies utilizing a gene fusion approach to analyze delivery of PrA to the vacuole in *Saccharomyces cerevisiae*.

Fusion of the gene encoding PrA, *PEP4*, to the *SUC2* gene, coding for the normally secreted yeast enzyme invertase, results in the production of a hybrid protein that retains invertase activity. The *PEP4-SUC2*-encoded hybrid proteins containing 76 amino acids or more from the N-terminus of PrA and lacking the invertase signal sequence are efficiently directed to the vacuole. The information necessary for directing PrA to the vacuole is then apparently contained within the N-terminal propeptide segment of the polypeptide.

This gene-fusion approach has also allowed for the selection of mutants that are defective in this delivery process. In these mutants, the PrA-invertase and CPY-invertase hybrid proteins, as well as the wild-type PrA and CPY proteins, are missorted and secreted to the cell surface. The mutations that lead to this effect are *trans*-acting and define at least 35 complementation groups. Further characterization of these mutants should enable us to develop a better molecular understanding of the mechanisms by which proteins are sorted and delivered to their ultimate subcellular compartments in eukaryotes.

Key words: carboxypeptidase Y, proteinase A, gene fusions, glycosylation, *Saccharomyces cerevisiae*

The vacuole of the yeast *Saccharomyces cerevisiae* is a lysosome-like organelle that contains a number of soluble hydrolytic glycoprotein enzymes [1,2]. These include proteinase A (PrA), proteinase B (PrB), carboxypeptidase Y (CPY), repressible alkaline phosphatase (rALPase), and RNase. There are also a variety of membrane-bound enzymes, including α-mannosidase, dipeptidylaminopeptidase B, a proton-translocating ATPase, and transport permeases [1,3–5] (Fig. 1). The primary role of the vacuole under normal growth conditions may be for storage of small molecules and osmoregulation. The level of vacuolar hydrolases varies with growth conditions and the stage of growth [nitrogen starvation, glucose depletion, and/or growth into stationary phase

Received June 15, 1987

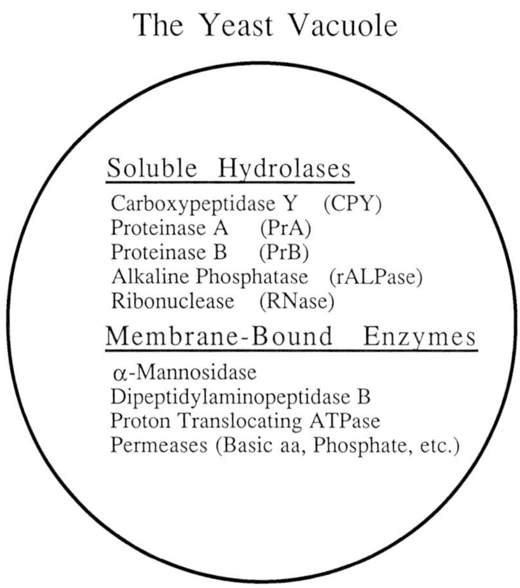

Fig. 1. The yeast vacuole contains a variety of soluble hydrolases, as well as several different membrane-bound proteins.

result in elevated activities for vacuolar proteins [6–8], suggesting a role in starvation-induced protein degradation and turnover. It is not yet known if the vacuole plays a specific role in the degradation of aberrant proteins. The vacuole normally accounts for approximately 25% of the cell volume. It is an acidic compartment and maintains the tonoplasm at a pH of approximately 5.8 [9].

Delivery of enzymes to the vacuole is mediated by a portion of the secretory pathway [10]. PrA, PrB, and CPY [11,12] have been shown to be synthesized as inactive proenzyme precursors in the endoplasmic reticulum (ER). All of the characterized vacuolar hydrolases appear to be mannoproteins and undergo dolichol-mediated core glycosylation in the ER [13,14]. Additional carbohydrate modification takes place in the Golgi complex. During transit through the Golgi, the proenzymes are separated from proteins destined for secretion. This segregation is thought to take place on the *trans* side of the Golgi. The presence of a receptor carrier has not been clearly demonstrated for any of the vacuolar proteins, although some evidence suggests that a receptor-mediated pathway may exist [15,16].

Just prior to or upon arrival in the vacuole, the proenzymes of CPY, PrA, and PrB are processed to their mature forms. Maturation is dependent on a functional *PEP4* (PrA) gene product [17]. The activities of RNase and alkaline phosphatase are also *PEP4*-dependent, suggesting that they too may be synthesized as precursors [17]. The activation of both proPrA and proCPY has been shown to require the removal of an amino terminal propeptide segment [17,18]. Processing of other vacuolar proteins is currently under investigation. The regulation of the maturation process is not understood, but it may involve a mechanism that is initiated by autocatalysis of propPrA [19]. Temperature-sensitive secretory mutants (*sec*) of yeast have been employed to trace the intracellular transit of CPY to the vacuole [10]. All *sec* mutants that, at the nonpermissive temperature, block secretory protein transit from the ER or from the Golgi also block

CPY delivery to the vacuole. However, *sec* mutants that block the transit of secretory proteins after the Golgi do not affect delivery of CPY to the vacuole [10]. This suggests that secretory proteins and vacuolar proteins travel together through the ER and Golgi before being sorted from one another for final delivery to their distinct cellular destinations (Fig. 2).

Unlike protein targeting to mammalian cell lysosomes, delivery to the vacuole is not dependent on glycosylation of the protein. In the presence of tunicamycin, a drug that blocks the synthesis of high-mannose core oligosaccharides, proCPY and alkaline phosphatase are delivered to the vacuole and matured [10,20,21]. This indicates that sorting of proenzymes to the vacuole may occur via direct recognition of a primary amino acid sequence or conformational feature of the polypeptide.

The genes coding for CPY (*PRC1*), PrA (*PEP4*), rALPase (*PHO8*), and PrB (*PRB1*) [6,16,18,19,22] have been cloned. Extensive analysis has been carried out on CPY to determine the factors that affect its targeting to the vacuole. CPY is initially synthesized in the ER as a 67,000-dalton precursor (p1) containing four Asn-linked core oligosaccharides. An N-terminal signal peptide of 20 amino acids serves as the signal

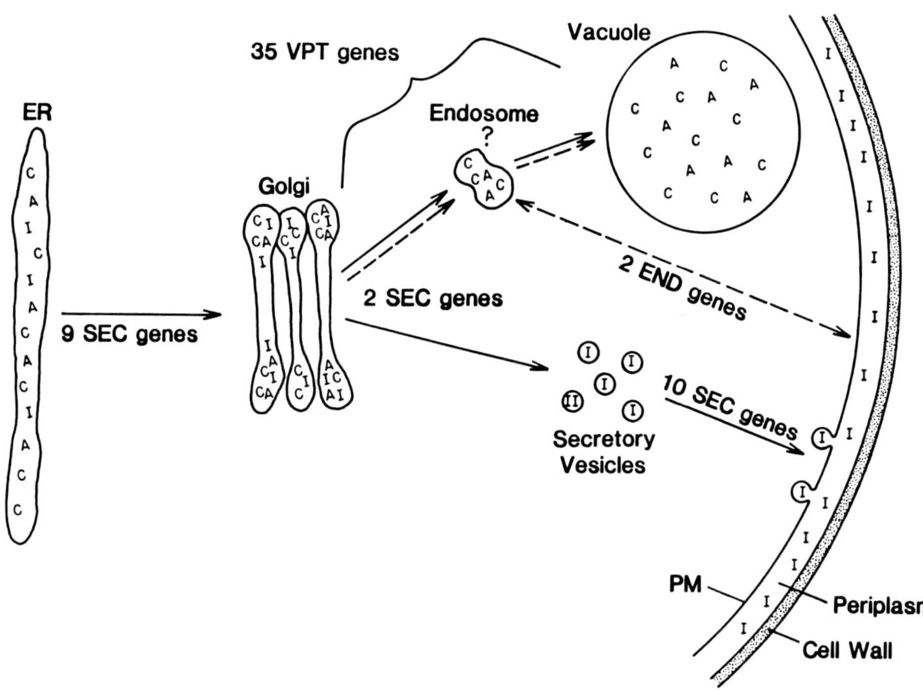

Fig. 2. The secretory enzyme invertase (I) and the vacuolar hydrolases carboxypeptidase Y (C) and proteinase (A) are synthesized in the cytoplasm and sequestered into the lumen of the endoplasmic reticulum (ER), where they are modified with N-glycosidically linked core oligosaccharides. The proteins then transit to the Golgi complex, where further glycosyl modification takes place. Invertase is then packaged in secretory vesicles that deliver this enzyme to the cell surface, while carboxypeptidase Y and proteinase A are targeted to the vacuole by a secretory vesicle-independent route. Different *SEC* genes have been identified and shown to code for functions required for interorganelle membrane and protein traffic as indicated [37]. In addition, two *END* genes coding for products involved in endocytosis are shown [38]. The *VPT* genes are largely uncharacterized but have been shown to affect delivery of vacuolar proteins [32, this report]. PM, plasma membrane.

to direct entry of the protein into the ER and is cleaved before transit out of the ER [23,24]. It is converted to a 69,000-dalton form (p2) in the Golgi before being delivered to the vacuole, where a 91-amino acid N-terminal propeptide is removed to produce mature CPY. The determinant that allows targeting of proCPY to the vacuole has been shown to reside in this N-terminal propeptide. Fusions of *PRC1* to *SUC2* [which codes for the secreted glycoprotein invertase (Inv)] showed that amino acids 21–50 of CPY were sufficient to direct a CPY-invertase hybrid protein to the vacuole [23]. Deletions of amino acids in the early part of the propeptide in the wild-type CPY protein resulted in missorting of proCPY [23,25]. The missorted proCPY is secreted into the periplasm and media.

Since many of the vacuolar proteases are synthesized as inactive precursors, it may be that the propeptide plays a role both in keeping the hydrolases inactive and in providing specific information that allows for efficient targeting to the vacuole. To further address the question of vacuolar targeting, we have employed the gene fusion approach in an attempt to map the recognition determinants in PrA that allow this protein to be efficiently delivered to the vacuole. Furthermore, the gene fusion approach has allowed us to devise selections for mutants that may be altered in cellular components that facilitate the sorting of proteins to the yeast vacuole.

PEP4-SUC2 GENE FUSIONS

Proteinase A is an aspartyl endoproteinase of 48,000–52,000 daltons containing two N-linked glycosidic side chains [11]. It is synthesized as an inactive precursor of 405 amino acids on the ER (p1) and undergoes further carbohydrate modification in the Golgi to produce a higher molecular weight (p2) form [11,14,18,19]. Upon delivery to the vacuole, an N-terminal propeptide is cleaved to generate the mature 329-amino acid enzyme with an apparent molecular weight of 42,000 daltons. The *PEP4* (PrA) gene of yeast was cloned by complementation in *PEP4* mutant yeast [15,19]. We have used the *PEP4* gene to construct fusions to the *SUC2* gene of yeast. The *SUC2* gene has also been cloned and sequenced [26,27]. We have used this gene to develop plasmid vectors (Fig. 3) that permit the simple construction of gene fusions to *SUC2* [28,29]. The *SUC2* gene in the fusion vectors has been deleted for the invertase signal sequence and the first two amino acids of the mature enzyme. These vectors can be maintained in both yeast and *E. coli*. In yeast, depending on the vector used, these plasmids are present in multiple copies per cell (2 μm DNA-based plasmid) or at approximately one copy per cell (*CEN4 ARS1*-based plasmid). Seven *PEP4-SUC2* gene fusions containing different sized amino terminal coding information from the *PEP4* gene fused to a constant fragment of the *SUC2* gene have been constructed (Fig. 4, unpublished results). Each of these gene fusions directs the synthesis of hybrid proteins in yeast that exhibit invertase activity. Immunoprecipitation of the PrA-Inv hybrid proteins from $^{35}SO_4^{2-}$ pulse-labeled yeast cells grown in the presence and absence of tunicamycin demonstrates that the hybrid proteins are glycosylated and therefore have presumably entered the yeast secretory pathway.

CELLULAR LOCATION OF PrA-INV HYBRID PROTEINS

The invertase enzyme activity associated with the *PEP4-SUC2*-encoded hybrid proteins has been used as a marker to detect the cellular location of these hybrid proteins.

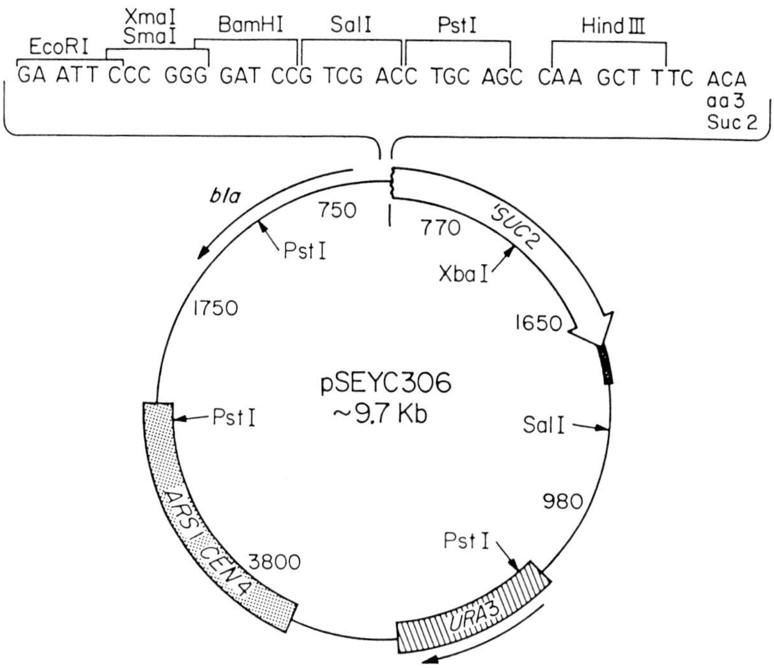

Fig. 3. Partial restriction map of the *SUC2* fusion vector pSEYC306. This vector contains the coding sequences for 511 amino acids of mature invertase. The coding sequences for the 19-amino acid invertase signal sequence and the first two amino acids of the mature enzyme have been deleted. A polylinker region containing unique *Eco*RI, *Sma*I, *Bam*HI, and *Hind*III restriction enzyme recognition sites was cloned at the 5′ end of the truncated *Suc2* gene. The vector pSEYC306 contains the *URA3* gene and the CEN4-ARS1 DNA sequences, allowing for the selection of this plasmid in yeast and its stable maintenance at a low copy number. An equivalent vector, pSEY304, replaces the CEN4-ARS1 DNA with the 2-μm circle origin of replication. This plasmid is maintained in multiple copies per cell. Both plasmids also contain the β-lactamase gene (bla) and a ColE1 origin of replication, which allows for their selection and maintenance in *E. coli*.

Enzyme assays with whole cells detect only the invertase activity that has been secreted from the cell [30]. Invertase assays using lysed cells permit detection of the total cellular enzyme activity. *PEP4-SUC2* fusions can be grouped into two classes, based on the location of the invertase activity generated by the hybrid proteins. The class I fusions represent those for which the majority of the invertase activity (> 90%) remains within the cell (Fig. 4, Table I). Yeast cells containing fusions with 76 amino acids or more of the N-terminus of preproPrA fused to invertase were found to express invertase activity that is retained within the cell (Table I). The PrA-Inv76 fusion (N-terminal 76 amino acids of preproPrA fused to invertase) contains the entire PrA signal sequence and propeptide region but none of the mature polypeptide. In contrast, *PEP4-SUC2* gene fusions expressing hybrid proteins that contain 61 amino acids or less of preproPrA (class II) secrete 65–95% of the hybrid proteins from the cell.

The shortest class I fusion contains the N-terminal 23 amino acids of preproPrA. The secretion of this hybrid protein from the cell indicates the presence of a functional signal sequence within the N-terminal 23 amino acids of the precursor protein. A signal sequence cleavage site is predicted to occur between amino acids 22 and 23 of the

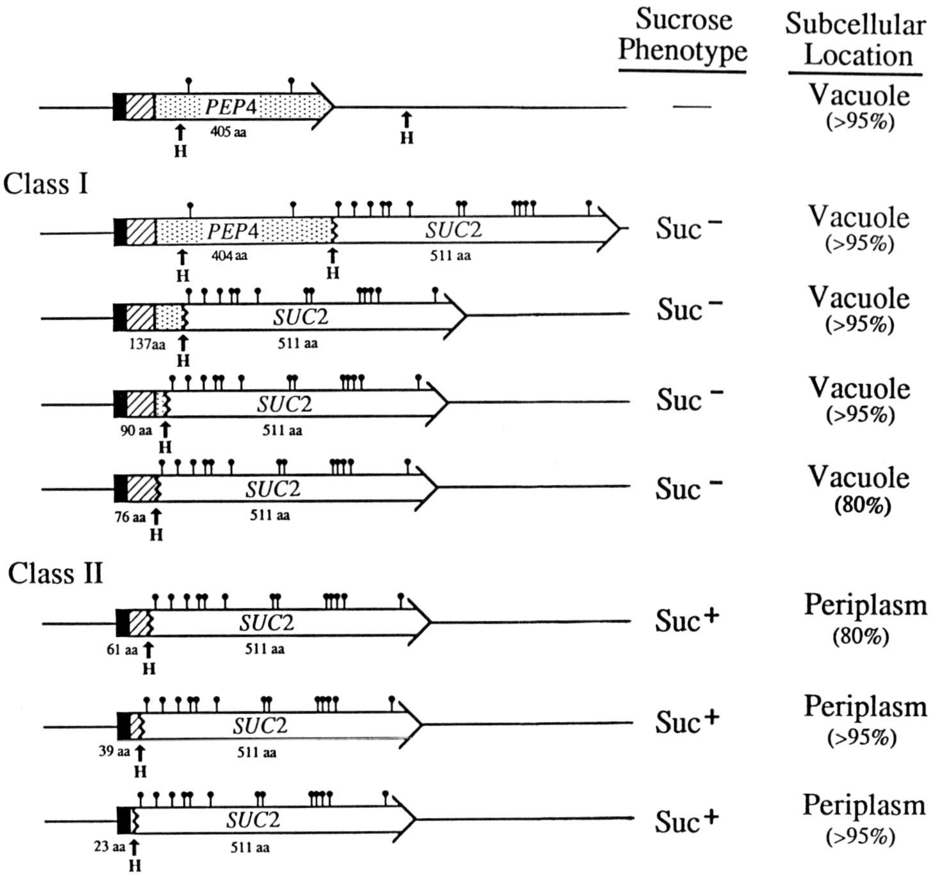

Fig. 4. Two classes of *PEP4-SUC2* gene fusions are shown. The *PEP4* gene is divided into the coding regions for: the presumed signal peptide (filled-in box), the propeptide (hatched box), and the mature PrA enzyme (stipled box). The approximate positions of the two asparagine-linked core oligosaccharides as determined from the amino acid sequence of the PrA protein [18,19] are indicated above the *PEP4* gene. The growth phenotype on sucrose of a yeast strain deleted for its normal suc loci and harboring each of the indicated gene constructs on a single-copy plasmid vector is shown. Cofractionation of the *PEP4* gene product and the hybrid protein products of each of the *PEP4-SUC2* gene fusions with isolated vacuoles is as indicated.

precursor protein, based on the rules of von Heijne [31]. In accordance with this prediction, the signal peptide appears to be cleaved in the PrA-Inv23 hybrid protein. The position of the vacuolar targeting signal within the PrA propeptide may be different from that for proCPY. A CPY-Inv50 fusion (the N-terminal 50 amino acids of preproCPY fused to invertase) resulted in approximately 90% localization of the hybrid protein in the vacuole, while the PrA-Inv61 fusion resulted in less than 10% delivery of the hybrid to the vacuole [23, unpublished results]. The proPrA sorting signal then is apparently located in a more distal part of the propeptide. It should be noted, however, that for both proCPY and proPrA, functional vacuolar targeting signals are present within their propeptides. This region does not contain sites for core oligosaccharide addition in either

TABLE I. Cellular Distribution of proPrA-Invertase Hybrid Proteins

Plasmids in strain SEY2108[a]	Total invertase enzyme activity (units/OD)	% Invertase activity	
		Extracellular	Intracellular
pP41-23	226	95	5
pP4I-39	264	92	8
pP4I-61	259	80	20
pP4I-76	426	10	90
pP4I-90	367	4	96
pP4I-137	442	3	97
pP4I-404	488	3	97

[a]Each of the *PEP4-SUC2* gene fusions was constructed in the fusion vector pSEYC306, with the exception of pP4I-404, which was constructed in the fusion vector pSEYC308.

protein. This supports the finding that delivery to the vacuole is not dependent on glycosylation [13,20,21].

To determine if the PrA-Inv hybrid proteins were being localized to the vacuole, cell extracts were fractionated on discontinuous Ficoll step-density gradients, and the isolated fractions were assayed for invertase activity [32]. For all of the class I hybrid proteins, the majority of the cell-associated invertase activity was found to cofractionate with the vacuolar enzyme α-mannosidase (Fig. 5). In contrast, the class II hybrid proteins show very little of the invertase activity in the vacuole fraction (< 10%).

DELETIONS IN WILD-TYPE PROTEINASE A

We have examined the kinetics of processing of wild-type PrA to determine the half-time of maturation. Pulse-chase labeling of yeast, followed by immunoprecipitation with antisera to PrA, showed a half-time of approximately 6 minutes for the processing of proPrA to the mature enzyme (Fig. 6). This is similar to the rate of maturation observed for proCPY [13,33].

To confirm the role of the amino terminal portion of proPrA in directing the vacuolar localization of PrA and to further define the location of the targeting signal, internal deletions in the wild-type PrA protein have been constructed. We have found that deletions of 15 amino acids or more in the propeptide region result in a dramatic decrease in the stability of proPrA. A deletion that removes the entire prosequence (amino acids 24–76), PrAΔpro, results in approximately 90% of the deleted protein being degraded within the ER (Fig. 6). Interestingly, the small amount of PrAΔpro that is stable (< 10%) undergoes a conversion to a higher-MW form. We believe this represents conversion from a p1 to a p2 form, corresponding to an ER → Golgi type glycosyl modification, since the terminal form comigrates with mature PrA. At least some fraction of this deleted protein then appears to undergo glycosyl processing in the Golgi complex. It should be noted, however, that this processing is much slower than that seen for the wild-type protein; it takes appoximately ten times longer to process PrAΔpro, compared with wild-type PrA. This instability of the mutant form of proPrA has thus far prevented us from analyzing the cellular localization of the deleted wild-type protein.

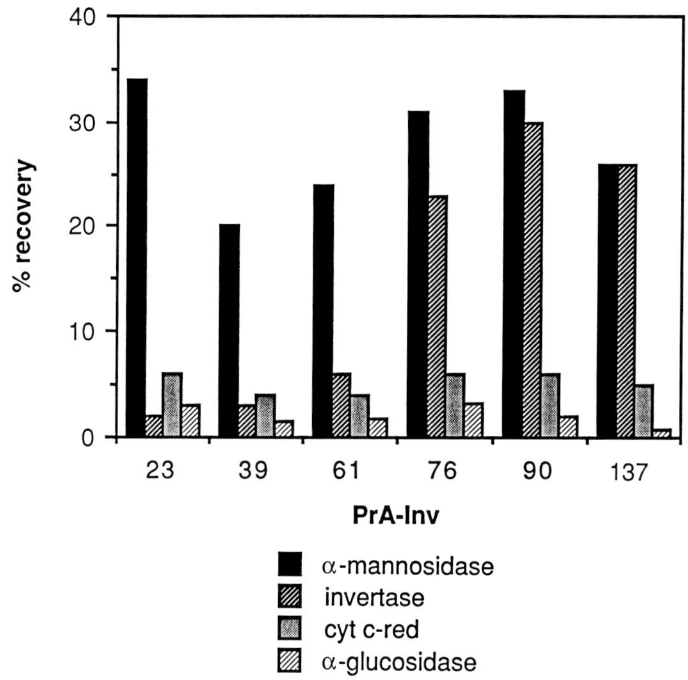

Fig. 5. Yeast strain SEY2108 containing a plasmid coding for the fusion protein PrA-Inv23, PrA-Inv39, PrA-Inv61, PrA-Inv76, PrA-Inv90, or PrA-Inv137 was fractionated and used to prepare purified vacuoles. Vacuoles were resolved on a discontinuous Ficoll step gradient. The percentage of recovery of α-mannosidase (vacuole membrane), invertase (PrA-invertase), α-glucosidase (cytoplasm), and cytochrome c reductase (ER) were determined by dividing the total enzyme activity recovered in the vacuole-enriched fraction by the total activity loaded on the gradient.

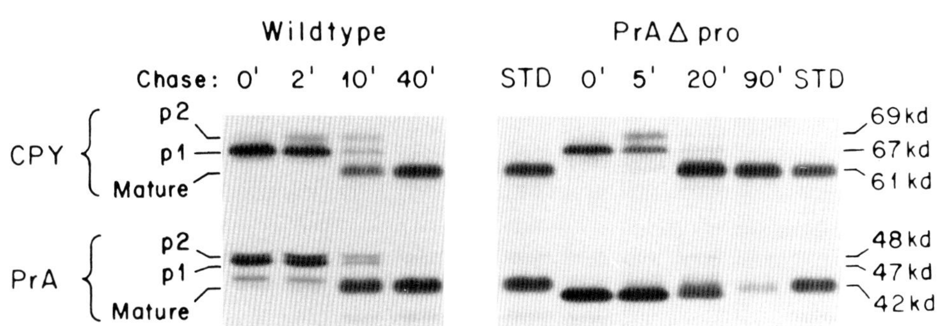

Fig. 6. Immunoprecipitation of wild-type proteinase A and proteinase A deleted for amino acids 24–76 (PrAΔpro). Yeast cells were labeled with $^{35}SO_4^{-2}$ for 5 minutes and chased with the addition of $MgSO_4$ to a final concentration of 50 mM for the indicated time. Radiolabeled proteins were precipitated with anti-proteinase A antiserum and analyzed by SDS-polyacrylamide gel electrophoresis and autoradiography. The approximate molecular weights of the proteins are indicated.

GLYCOSYLATION OF HYBRID PROTEINS

All vacuolar proteases characterized to date undergo only limited mannosyl processing prior to delivery to the vacuole. N-linked core oligosaccharides are added in the ER by a dolichol-mediated process. During transit through the ER and Golgi complex, the core groups are trimmed and modified, resulting in a final glycosidic side chain containing 10–18 mannose residues [34]. This is contrasted with the secreted enzyme invertase, which undergoes extensive glycosyl modification; 100–150 mannose residues are added to some of the core oligosaccharides, resulting in a hyperglycosylated protein with extensive outer chain carbohydrates [27,35]. This hyperglycosylated invertase migrates as a broad smear on SDS polyacrylamide gels due to its being a heterogeneous mixture of highly glycosylated molecules (Fig. 7). In contrast, the limited extension of CPY oligosaccharide side chains results in a protein that migrates as a discrete species on polyacrylamide gels (Fig. 7). The hybrid CPY-Inv proteins contain all of the oligosaccharide addition sites present in wild-type invertase. The CPY-Inv50 hybrid protein does not contain any of the CPY oligosaccharide addition sites, while the CPY-Inv433 hybrid has three of the four sites present in wild-type CPY (Fig. 8). Immunoprecipitation of wild-type or hybrid proteins in a *sec18* strain at the nonpermissive temperature shows the proteins with core oligosaccharides added as they are found in the ER [23] (Fig. 7). The apparent MW corresponds to the polypeptide MW plus the addition of core oligosaccharides at the predicted oligosaccharide addition sites. Immunoprecipitation of invertase from wild-type yeast shows the broad smear typical for this protein. The CPY-Inv50 hybrid protein migrated in a similar fashion, suggesting that this hybrid protein is modified in a manner characteristic of wild-type invertase. The CPY-Inv50 hybrid protein must then enter Golgi compartments containing the mannosyl transferases responsible for outer chain mannose addition, even though it is eventually delivered to the vacuole. Then the sorting of vacuolar and secretory proteins apparently occurs after this point in the Golgi. In addition, delivery to the vacuole does not, in itself, determine the extent of oligosaccharide modification of the hybrid protein. In contrast to the result with CPY-Inv50, the CPY-Inv433 hybrid protein migrated as a distinct species on SDS-polyacrylamide gels [23] (Fig. 7), indicating that it received only limited mannosyl modification. This led to the proposal that a determinant mapping distal to the vacuolar sorting signal could specify, in a *cis*-dominant manner, the extent to which CPY and CPY-Inv hybrid proteins are glycosylated in the Golgi [23]. A similar situation appears to exist for the PrA-Inv hybrid proteins. Fusions of 137 amino acids or less of PrA to invertase result in hybrid proteins that show the hyperglycosylation typical of invertase. In contrast, a fusion of almost the entire PrA protein to invertase, PrA-Inv404, appears to receive more limited glycosyl modificaton. We are currently constructing additional CPY-Inv and PrA-Inv fusions in an attempt to more carefully map the region responsible for regulation of carbohydrate modification.

MUTANTS DEFECTIVE IN VACUOLAR PROTEIN TARGETING (*vpt*)

Yeast cells deleted for the chromosomal *SUC2* locus are unable to utilize sucrose as the sole fermentable carbon source (Suc^-). When these cells harbor a single-copy plasmid coding for CPY-Inv or PrA-Inv hybrid proteins that are secreted from the cell (class II fusions), they become phenotypically Suc^+ (Fig. 4). When they contain single-copy plasmids coding for hybrid proteins that are efficiently directed to the vacuole,

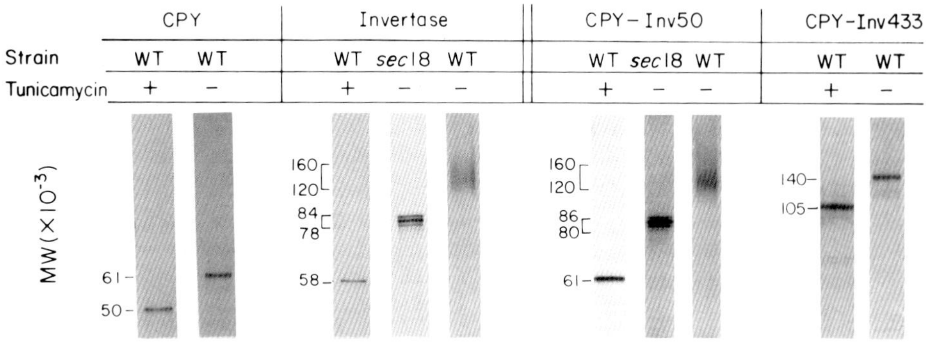

Fig. 7. Immunoprecipitation of carboxypeptidase Y, invertase, and *PRC1-SUC2* hybrid proteins in a wild-type (SEY2108) and a *sec18* mutant (SEY5184; transit from the ER is blocked at the nonpermissive temperature) yeast strain. Wild-type cells expressing the indicated proteins were labeled for 30 minutes at 30°C. Where indicated, tunicamycin (final concentration of 20 μg/ml) was added 10 minutes before the addition of label to inhibit glycosylation. *sec18* cells were grown at 22°C and shifted to 32°C 5 minutes before the addition of label. Immunoprecipitations were carried out using anti-carboxypeptidase Y or anti-invertase antiserum. The approximate molecular weights are indicated.

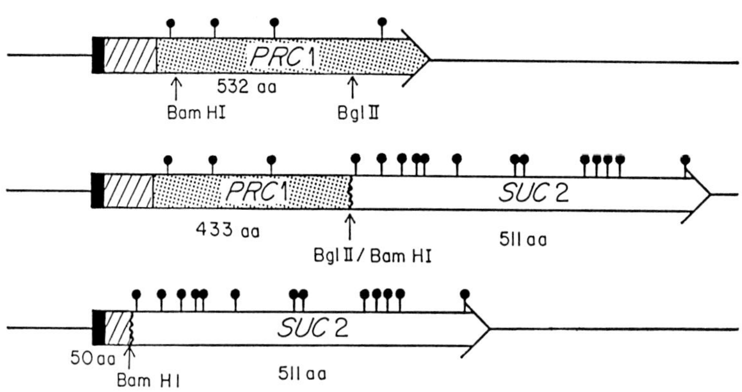

Fig. 8. The *PRC1* gene and two *PRC1-SUC2* gene fusions. The designations for the *PRC1* coding regions are the same as those used in Figure 4. The approximate positions of the asparagine-linked core oligosaccharide addition sites as determined from the CPY and invertase amino acid sequences are indicated.

however, the cells are Suc$^-$, presumably because sucrose cannot gain access to the hybrid protein-derived invertase activity in the vacuole. We have taken advantage of this phenotype to select for mutants (Suc$^+$) defective in *v*acuolar *p*rotein *t*argeting (*vpt*) of the hybrid proteins [32, unpublished results]. Analysis of such mutants has demonstrated that as much as 90–100% of the total invertase activity expressed in these mutants is secreted from the cell. The mutants are not, however, altered in their total level of cellular invertase activity. Therefore secretion of CPY-Inv or PrA-Inv in these mutants is not simply the result of overexpression of the hybrid proteins, nor is it simply the result of the mutants being prone to cell lysis, as we do not detect elevated levels of cytoplasmic enzyme activities in extracellular fractions from these mutants.

We have found that almost none of the Suc$^+$ mutants contain mutations linked to the gene fusions. Instead, they carry mutations in other nuclear genes. Complementation analysis has revealed that the mutations are almost all recessive to wild-type and define at least 35 complementation groups. Presumably the genes altered in these mutants code for proteins that function in regulating protein traffic through the yeast secretory pathway. Protein secretion does not appear to be altered in the mutants, however, suggesting that they may be specifically defective either in the initial sorting of the hybrid proteins or in their delivery from the Golgi to the vacuole.

The vacuolar protein targeting defect observed in the *vpt* mutants is not restricted to CPY-Inv and PrA-Inv hybrid proteins, since a fusion-independent mislocalization of wild-type CPY and PrA is also detected in these strains (Fig. 9). Radiolabeling experiments demonstrate that *vpt* mutants exhibit major defects in the conversion of proCPY and ProPrA to their mature forms. In some *vpt* strains, there is an accumulation of the proenzyme forms, and only a small amount of mature CPY or PrA is detected (Fig. 9). Fractionation of *vpt* mutant strains has shown that both the mature and proenzyme forms of CPY and PrA are secreted from the cell [32, unpublished results]. There is an apparently equal distribution of the secreted vacuolar proteins between the periplasmic and medium fractions, and the secretion shows kinetics comparable to those for normally secreted yeast glycoproteins.

We have grouped the *vpt* mutant strains into three classes, based on morphological differences. Class A *vpt* strains contain a normal vacuolar compartment as detected by

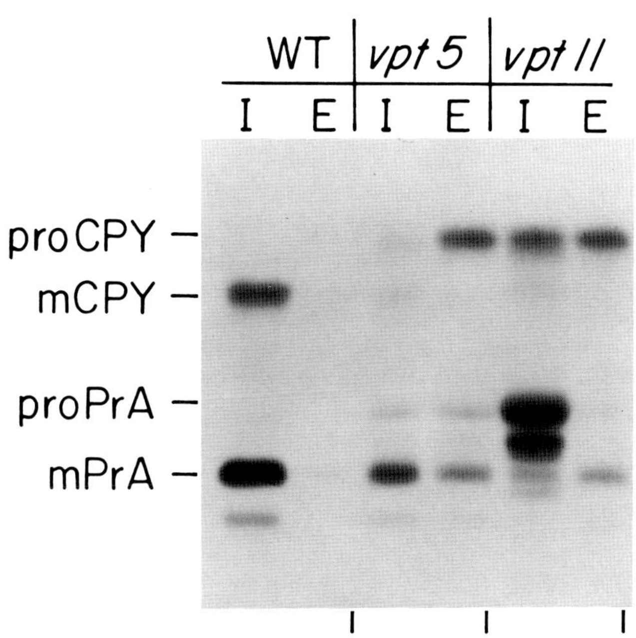

Fig. 9. Immunoprecipitation of CPY and PrA in wild-type and *vpt* mutant yeast strains. Cells were labeled with $^{35}SO_4^{-2}$ for 20 minutes and chased with the addition of $MgSO_4$ (final concentration 50 mM) for 20 minutes in the presence of BSA (1 mg/ml final concentration) to reduce nonspecific proteolysis. Cells were then enzymatically converted to spheroplast (I) and periplasm plus medium (E) fractions. Radiolabeled proteins were coimmunoprecipitated with anti-carboxypeptidase Y and anti-proteinase A antiserum.

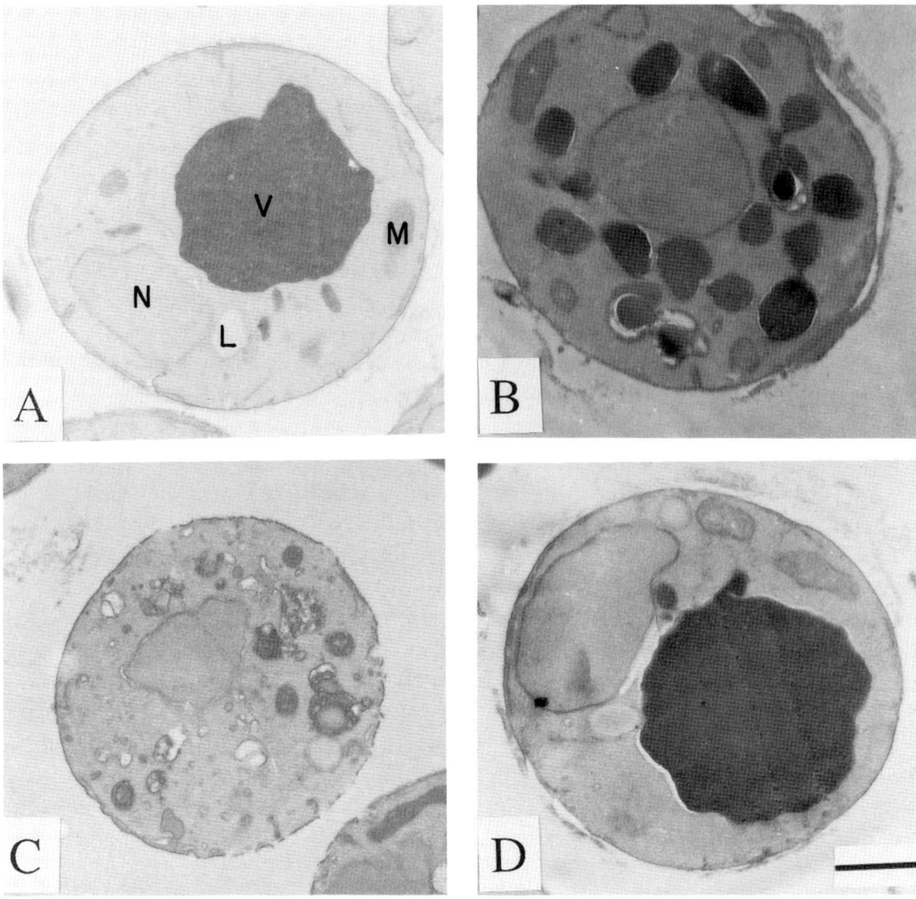

Fig. 10. Morphology of *vpt* mutant strains. *vpt* mutant strains with Class A **(A)**, Class B **(B)**, or Class C **(C)** morphology and the isogenic wild-type strain SEY6210 **(D)** were prepared for electron microscopy essentially as described by Byers [39]. Cells were further processed by the reduced osmium-thiocarbohydrazide-reduced osmium (R-OTO) procedure [40]. Cells were then washed, dehydrated with ethanol, and embedded in LR white. Thin sections were stained with lead citrate [41]. V, vacuole; M, mitochondria; N, nucleus; L, lipid droplet. The bar represents 1 μm.

Normarski optics or electron microscopy [unpublished results] (Fig. 10). Most of the *vpt* strains are characterized as having a class A morphology. In contrast, the class B *vpt* mutants lack the large vacuole seen in wild-type yeast. They do contain, however, an average of 30–50 smaller membrane-bounded compartments (Fig. 10). It is not yet known if these compartments are related to the vacuole, but histochemical staining indicates that they may correspond to a fragmented vacuole. The class C *vpt* strains show an even greater morphological defect in that they contain no identifable vacuole-like compartment and accumulate vesicles and abnormal membrane-enclosed structures (Fig. 10). These mutants also tend to show the most severe growth defects.

The *vpt* mutants then show defects in delivery to the vacuole of at least two vacuolar enzymes. Our preliminary data also indicate that an additional vacuolar enzyme, repressible alkaline phosphatase, is subject to missorting in at least some of the *vpt* mutant strains. It appears, therefore, that a common mechanism may be responsible for proper targeting of several vacuolar proteins to their correct organelle destination. While the *vpt* mutants are pleiotropic in their effect on certain vacuolar proteins (in particular the lumenal enzymes), others, such as the membrane-bound α-mannosidase, appear to be delivered to the vacuole in most of the *vpt* mutants. Those proteins that show normal targeting in the *vpt* strains may be delivered by a pathway that is separate from that used for CPY and PrA. The *vpt* mutants should enable us to define some of the components that function in the specific sorting and delivery of these vacuolar proteins. Tom Stevens and his coworkers at the University of Oregon have also isolated mutants that exhibit defects in protein delivery to the vacuole by a different approach [36]. Comparisons between these two sets of mutants has recently been initiated. Together these studies suggest that protein segregation to the vacuole, like other events in the secretory pathway [37], is a complex process requiring the participation of a large number of gene functions.

CONCLUSIONS

We have used a gene fusion approach to identify a region of carboxypeptidase Y and proteinase A that allows these proteins to be targeted efficiently to the vacuole. Our analysis of PrA-Inv hybrid proteins indicates that a vacuolar sorting signal is contained within the PrA propeptide. A similar result has been found for CPY [23]. We have compared the amino acid sequences of preproPrA and preproCPY to identify common features. At present we have been unable to find any obvious primary sequence homology in these vacuolar sorting signals. Potentially, different recognition systems may participate in the sorting of these proteins or, more likely, some other features of these sequences, such as their secondary or tertiary structure, may play a significant role in the recognition event. The sorting signal of proCPY has been best characterized, and it is still only mapped to a 20–30-amino acid region of the propeptide. The important characteristics of this signal are not at all understood. In this sense it is not surprising that we cannot identify common features, if any, between the proPrA and the proCPY sorting signals. We have recently begun to analyze the location of the targeting determinant of the vacuolar enzyme alkaline phosphatase. Our preliminary data indicate that an N-terminal segment of this protein contains information that directs the enzyme to the vacuole. Further analysis of this and other vacuolar enzymes, coupled with a more detailed analysis of the targeting signals in CPY and PrA, should provide further insight into the nature of the vacuolar sorting signal. We also have been able to genetically define a number of genes, the products of which appear to play a role in vacuolar protein targeting. Our initial results indicate that this is a complex process requiring the participation of a large number of gene products. It is hoped that by continued genetic analysis of yeast vacuole assembly, coupled with biochemical characterization of the gene products that function in this process, we will be able to develop a better molecular understanding of the mechanisms by which proteins are sorted and delivered to their ultimate subcellular compartments in eukaryotes.

ACKNOWLEDGMENTS

This work was supported by grants from the National Institutes of Health (GM 32703) and the Chicago Community Trust/Searle Scholars Program. D.J.K. was supported by a Helen Hay Whitney Postdoctoral Research Fellowship. L.M.B. and J.S.R.

were supported by graduate research fellowships from the National Science Foundation and Evelyn Sharp fund, respectively.

REFERENCES

1. Wiemken A, Schellenberg M, Urech K: Arch Microbiol 123:23, 1979.
2. Lenney J, Matile P, Wiemken A, Schellenberg M, Meyer J: Biochem Biophys Res Commun 50:1378, 1974.
3. Kakinuma Y, Ohsumi Y, Anraku Y: J Biol Chem 256:10859, 1981.
4. Boller Th, Dürr M, Wiemken A: Eur J Biochem 54:81, 1975.
5. Bordallo C, Schwencke J, Suarez Rendueles MS: FEBS Lett 173:199, 1984.
6. Moehle CM, Aynardi MW, Kolodny MR, Park FJ, Jones EW: Genetics 115:255, 1987.
7. Saheki T, Holzer H: Biochem Biophys Acta 384:203, 975.
8. Hansen R, Switzer R, Hinze H, Holzer H: Biochem Biophys Acta 196:103, 1977.
9. Navon G, Shulman RG, Yamane T, Eccleshall TR, Lam K-B: Biochemistry 18:4487, 1979.
10. Stevens T, Esmon B, Schekman R: Cell 30:439, 1972.
11. Mechler B, Müller M, Müller H, Meussdoerffer G, Wolf DH: J Biol Chem 257:11203, 1982.
12. Hasilik A, Tanner W: Eur J Biochem 85:599, 1978.
13. Hasilik A, Tanner, W: Eur J Biochem 91:567, 1978.
14. Zubenko GS, Park FJ, Jones EW: Proc Natl Acad Sci USA 80:510, 1983.
15. Rothman JH, Hunter CP, Valls LA, Stevens TH: Proc Natl Acad Sci USA 83:3248, 1986.
16. Stevens TH, Rothman JH, Payne GS, Schekman R: J Cell Biol 102:1551, 1986.
17. Hemmings BA, Zubenko GS, Hasilik A, Jones EW: Proc Natl Acad Sci USA 78:435, 1981.
18. Ammerer G, Hunter CP, Rothman JH, Saari GC, Valls LA, Stevens TH: Mol Cell Biol 6:2490, 1986.
19. Woolford CA, Daniels LB, Park FJ, Jones EW, van Arsdell JN, Innis MA: Mol Cell Biol 6:2500, 1986.
20. Schwaiger H, Hasilik A, von Figura K, Wiemken A, Tanner W: Biochem Biophys Res Commun 104:950, 1982.
21. Clark DW, Tkacz JS, Lampen JO: J Bacteriol 152:865, 1982.
22. Kaneko Y, Tamai Y, Toh-e A, Oshima Y: Mol Cell Biol 5:248, 1985.
23. Johnson LM, Bankaitis VA, Emr SD: Cell 48:875, 1987.
24. Blachly-Dyson E, Stevens TH: J Cell Biol 104:1183, 1987.
25. Valls LA, Hunter CP, Rothman JH, Stevens TH: Cell 48:887, 1987.
26. Carlson M, Botstein D: Cell 28:145, 1982.
27. Taussig R, Carlson M: Nucleic Acids Res 11:1943, 1983.
28. Emr SD, Schekman R, Flessel MC, Thorner J: Proc Natl Acad Sci USA 80:7080, 1983.
29. Emr SD, Vassarotti A, Garrett J, Geller BL, Takeda M, Douglas M: J Cell Biol 102:523, 1985.
30. Novick P, Field C, Schekman R: Cell 21:205, 1980.
31. von Heijne G: Eur J Biochem 133:17, 1983.
32. Bankaitis VA, Johnson LM, Emr SD: Proc Natl Acad Sci USA 83:9075, 1986.
33. Hasilik A, Tanner W: Biochem Biophys Res Commun 72:1430, 1976.
34. Trimble RB, Maley F, Chu FK: J Biol Chem 258:2562, 1983.
35. Ballou C: In Strathern J, Jones E, Broach J (eds): "The Molecular Biology of the Yeast Saccharomyces: Metabolism and Gene Expression." New York: Cold Spring Harbor Laboratory, 1982, pp 335–360.
36. Rothman JH, Stevens TH: Cell 47:1041, 1986.
37. Novick P, Ferro S, Schekman R: Cell 25:461, 1981.
38. Chvatchko Y, Howald I, Riezman H: Cell 46:355, 1986.
39. Byers B, Goetsch L: J Bacteriol 124:511, 1975.
40. Willingham MC, Rutherford AV: J Histochem Cytochem 32:455, 1984.
41. Reynolds ES: J Cell Biol 17:208, 1963.

Molecular Biology of Intracellular Protein
Sorting and Organelle Assembly 187–192 (1988)

Is Nuclear Protein Localization Receptor-Mediated?

Michael N. Hall and Sharon R. Fried

Department of Biochemistry and Biophysics, University of California, San Francisco, San Francisco, California 94143

The identification and characterization of nuclear localization signals in the yeast protein α2 are reviewed. Evidence that these signals, and signals in other nuclear proteins, bind a receptor, as proposed by a selective uptake model of nuclear localization, is discussed.

Key words: yeast, nucleus, signal, *MATα2*

Nuclear localization signals have recently been identified in several nuclear proteins [for review see 1,2]. The discovery of these signals has changed perceptions of how specific proteins accumulate in the nucleus. Nuclear proteins were thought to accumulate in the nucleus by first passively diffusing through pores in the nuclear envelope and then being selectively retained by binding to a nondiffusible intranuclear substrate [3]. The properties of the recently discovered nuclear localization signals are not consistent with this selective retention model and have consequently led to acceptance of an alternative model, selective uptake (Fig. 1) [4]. According to a selective uptake model, nuclear proteins are localized by an active process involving an interaction between a nuclear localization signal and a yet-to-be identified receptor. Here we review the identification and characterization of nuclear localization signals in the yeast protein α2 and discuss evidence that these signals, and signals in other nuclear proteins, do indeed bind a receptor as proposed by the selective uptake model.

TWO DISTINCT NUCLEAR LOCALIZATION SIGNALS

The α2 protein, the product of the *MATα2* gene, is a DNA-binding protein that regulates cell type in the yeast *Saccharomyces cerevisiae*. It represses transcription of a group of cell type-specific genes by binding to an operator located upstream of each target gene [5]. The α2 monomer, consisting of 210 amino acids, has a molecular weight of approximately 25 kD.

To identify a nuclear localization signal in α2, an extensive series of gene fusions encoding hybrid proteins containing different portions of the α2 protein fused to β-

Received June 18, 1987.

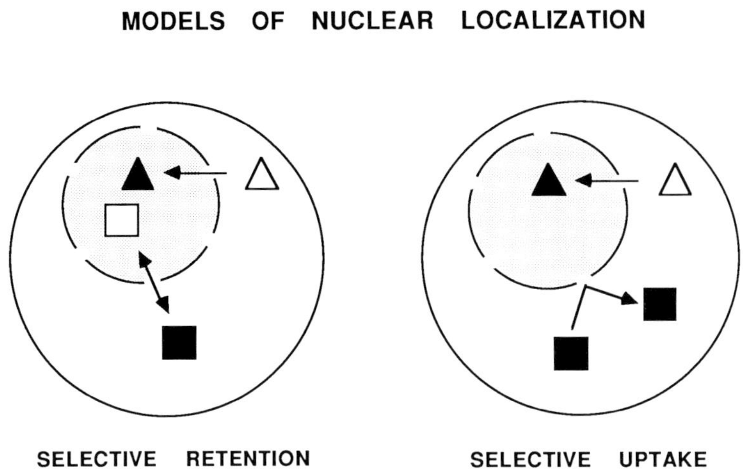

Fig. 1. According to a selective retention model, all proteins diffuse into the nucleus, but only those that bind an intranuclear nondiffusible substrate are retained. According to a selective uptake model, nuclear proteins are selectively translocated across the nuclear envelope. The passage of proteins into the nucleus is through pore complexes in the nuclear envelope. Triangles represent nuclear proteins; squares represent cytoplasmic proteins.

galactosidase was constructed [6,7] (Fig. 2). The cellular location of the different hybrid proteins was determined by indirect immunofluorescence on whole fixed cells. The results of this study are summarized in Figure 2. According to these results, the α2 protein has two signals, either one of which is sufficient for directing β-galactosidase to the nucleus. One of the signals is in the N-terminal 13 amino acids. This follows from the observation that a hybrid protein containing only these 13 amino acids of α2 fused to β-galactosidase is targeted to the nucleus [6]. The second signal resides, at least in part, in amino acids 141 through 152. This is based on the observation that a hybrid protein missing amino acids 4 through 140 is localized to the nucleus, whereas a protein missing 12 more amino acids, 4 through 152, is dispersed throughout the cell. The existence of two signals, one at the N-terminus and one between amino acids 140 and 153 is further supported by the finding that a protein containing two small deletions that eliminate only these two regions, does not accumulate in the nucleus [7].

The two regions of α2 shown to contain a localization determinant do not resemble each other, nor do they exhibit any homology with the well-characterized SV40 T antigen signal [8]. However, comparison of amino acid sequences of other yeast proteins with the N-terminal 13 amino acids of α2 first revealed a sequence that could be important for nuclear targeting, Lys_3-Ile-Pro-Ile-Lys_7. A similar sequence of two positively charged amino acids flanking three hydrophobic residues, one of which is proline, is present in several other yeast nuclear proteins (Fig. 3) [6] and is not present in any yeast cytoplasmic protein currently known. The α2 N-terminal sequence was subsequently shown to indeed be important as a nuclear localization determinant by deletion analysis (Fig. 2) [7]. Regions of influenza virus nucleoprotein [9] and *Xenopus* nucleoplasmin [10] that have been shown to mediate nuclear accumulation contain a similar sequence. The sequence of the internal region of α2 shown to contain a localization signal, amino acids 141 through 152, is Val_{141}-Arg-Ile-Leu-Glu-Ser-Trp-Phe-Ala-Lys-Asn-Ile_{152}.

The α2 protein is not unique in having two localization signals. Influenza virus

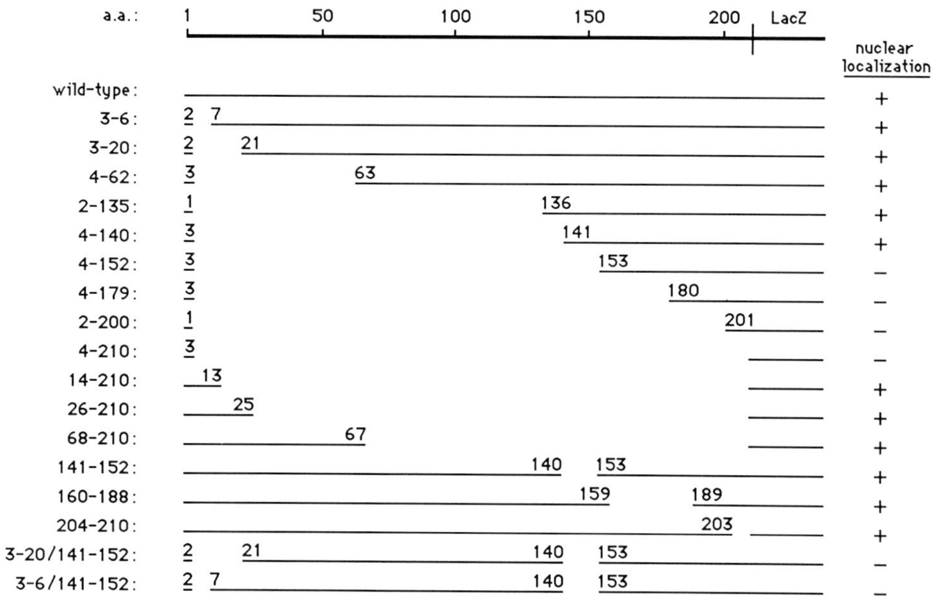

Fig. 2. Properties of α2-LacZ hybrid proteins. The hybrid proteins contain different portions of the α2 protein at the N-terminus and a constant, enzymatically active portion of *E. coli* β-galactosidase at the C-terminus. The wild-type hybrid protein contains all 210 amino acids of α2. The numbers in the left-hand column refer to the amino acids of α2 that are missing. The number at each deletion endpoint refers to the α2 amino acid at that position. β-galactosidase, consisting of 1,016 amino acids, is not drawn to scale. Nuclear localization was determined by indirect immunofluorescence.

nucleoprotein [9], polyoma virus T antigen [11], SV40 T antigen [12], and glucocorticoid receptor [13] are also known or suspected to have two localization determinants. Two localization signals could be a common feature of the process by which proteins are delivered to and translocated across the nuclear envelope.

ROLE OF THE NUCLEAR LOCALIZATION SIGNALS

What They Are Not

Two possible roles that could be ascribed to the α2 nuclear localization signals to mediate the observed nuclear accumulation have been ruled out. Neither signal in α2 confers nuclear localization through DNA binding or α2-subunit multimerization.

According to the passive diffusion and selective retention model of nuclear localization, the function of the signals would be to bind DNA, the intranuclear substrate of α2. This prediction was tested by assaying the different hybrid proteins for nonspecific DNA binding. Determination of nonspecific DNA binding is appropriate for this study, even though α2 is a specific DNA-binding protein, since specific DNA binding alone would not be sufficient to retain α2 in the nucleus due to the relative scarcity of specific binding sites. The results [14, unpublished data] of this test reveal that there is no correlation between nuclear localization of the hyrid proteins and an ability to bind DNA in vitro. This indicates that DNA binding is neither necessary nor sufficient for nuclear localization, as has been observed in other systems [15,16]. Assuming that α2 does not have a second nondiffusible intranuclear substrate that could retain α2 in the absence of

	+	HYDROPHOBIC	+		
MET ASN	LYS 3	ILE PRO ILE	LYS 7	ASP LEU LEU ASN PRO GLN	α2 (210A.A.)
MET PHE THR SER	LYS 5	PRO ALA PHE	LYS 9	ILE LYS ASN LYS ALA SER	α1 (175A.A.)
PRO ALA GLU LYS	LYS 18	PRO ALA ALA	LYS 22	LYS THR SER THR SER THR	H2B-1 (131A.A.)
GLN GLY ILE THR	LYS 32	PRO ALA ILE	ARG 36	ARG LEU ALA ARG ARG GLY	H4 (103A.A.)
ASN LYS LYS THR	ARG 79	ILE ILE PRO	ARG 83	HIS LEU GLN LEU ALA ILE	H2A (132A.A.)
SER VAL TYR GLU	LYS 411	PHE ALA PRO	LYS 415	GLY LYS GLN LEU SER MET	RAD52 (504A.A.)
HIS ALA PRO LEU	LYS 112	PRO VAL VAL	ARG 116	LYS LYS PRO GLU PRO ILE	SPT2 (333A.A.)

Fig. 3. Sequence of the N-terminal nuclear localization determinant of α2 and similar sequences found in other known or presumed nuclear proteins α1 [26], H2B-1 [27], H4 [28], H2A [29], RAD52 [30], and SPT2 [31]. The number underneath an amino acid indicates the position of the amino acid within the respective protein. Numbers in parentheses indicate the total number of amino acids in the designated protein.

DNA binding, the above results argue against selective retention as the mechanism of nuclear localization.

Studies on *Xenopus* nucleoplasmin, a nuclear protein of five identical subunits, suggest that a localization defective subunit can "piggyback" into the nucleus by multimerizing with a wild-type subunit [17]. Accordingly, one of the localization determinants in α2 could be a multimerization domain misconstrued as a localization signal. This possibility has been ruled out in the case of α2, since none of the proteins that contain only one of the two signals requires the presence of wild-type α2 to be delivered to the nucleus [7].

What They May Be

Our working model for the function of the signals in α2 is that these signals bind receptors that mediate selective uptake of nuclear proteins. Properties of certain mutant hybrid proteins and results obtained in other systems support this model.

Properties of the α2 proteins that are missing one of the two localization signals are consistent with the notion that the signals interact with receptors. Hybrid proteins that are missing the internal signal are lethal when overproduced [6] and accumulate in the nuclear envelope [7]. The simplest interpretation of these findings is that these proteins are able to bind a receptor on the nuclear envelope but are unable to translocate across the envelope. Overproduction lethality could be due to exclusion of essential proteins from the nucleus as a result of receptors being occupied. Accordingly, the N-terminal signal binds a surface receptor, and the internal signal is subsequently required for translocation across the nuclear envelope, perhaps by binding a second receptor. A two-step localization pathway such as this could account for why α2 and other nuclear proteins have two localization signals. The α2 proteins that are missing the N-terminal signal are localized to the nucleus, as determined by immunofluorescence [7]. However, these mutant proteins lack repressor activity in vivo, despite still being able to bind DNA specifically in vitro [14]. Loss of repressor activity despite retention of DNA-binding ability can be explained by proposing that elimination of the N-terminal signal does indeed affect nuclear localization, but in a manner not detectable by indirect immunofluorescence, e.g., the rate of localization could be altered. The interaction with the surface receptor by the N-terminal signal, as proposed above, is perhaps not essential for nuclear localization as determined by indirect immunofluorescence but is essential as determined by an assay of function.

Explicit or implict in a receptor-mediated uptake model such as the admittedly speculative model proposed above for α2 are: 1) an energy requirement; 2) rates suggestive of mediated uptake; 3) saturability; 4) competition; and 5) specificity. Observations made in other laboratories suggest that the process of nuclear protein localization does indeed have these features. ATP has been shown to be required for accumulation of nucleoplasmin in *Xenopus* nuclei [18,19]. Mathematical analysis of data on uptake of nuclear proteins indicates that the observed rates of nuclear accumulation are too great to be accounted for by a nonmediated diffusion and selective retention mechanism [10,20]. Studies on nuclear uptake of BSA cross-linked to a synthetic peptide containing the SV40 T antigen nuclear localization signal have shown that nuclear accumulation is saturable and can be inhibited by excess free peptide [21,22]. The nuclear accumulation of signal peptide-linked BSA injected into *Xenopus* oocytes exhibits saturation kinetics, and the rate of accumulation is diminished by coinjection of free peptide. Similarly, nuclear accumulation of injected histones is saturable, and histones of different species can compete with each other for accumulation in the nucleus [23]. All these observations taken together suggest strongly that uptake of protein by the nucleus is a receptor-mediated, active process.

THE RECEPTOR

As described above, several lines of evidence suggest that nuclear localization is receptor-mediated. However, confirmation of the involvement of a receptor will require identification of the actual receptor. Perhaps the most logical place to search for the receptor is the nuclear pore complex. Proteins are known to enter the nucleus through these complexes in the nuclear envelope [24]. Nuclear uptake is inhibited by a lectin that binds nuclear pores [25]. In addition, the accumulation of mutant α2 in the nuclear envelope, as described above, appears to be at discrete sites, assumed to be nuclear pores. Until a receptor is identified, the question of whether or not nuclear localization is a receptor-mediated process remains open.

ACKNOWLEDGMENTS

This work was supported by a research grant (GM35284) from the National Institute of General Medical Sciences awared to M.N.H.

REFERENCES

1. Dingwall C, Laskey RA: Annu Rev Cell Biol 2:367–390, 1986.
2. Silver PA, Hall MN: In Das RC, Robbins PW (eds): ''Protein Transport and Organelle Biogenesis.'' Orlando, Florida: Academic Press Inc., in press, 1987.
3. Bonner WM: In Busch H (ed): ''The Cell Nucleus.'' New York: Academic Press Inc., 1978, pp. 97–148.
4. De Robertis EM: Cell 32:1021–1025, 1983.
5. Johnson AD, Herskowitz I: Cell 42:237–247, 1985.
6. Hall MN, Hereford L, Herskowitz I: Cell 36:1057–1065, 1984.
7. Hall MN, Craik C, Mullenbach G, Hiraoka Y: in preparation, 1988.
8. Kalderon D, Roberts BL, Richardson WD, Smith AE: Cell 39:499–509, 1984.
9. Davey J, Dimmock NJ, Colman A: Cell 40:667–675, 1985.
10. Dingwall C, Burglin TR, Kearsey SE, Dilworth S, Laskey RA: Peters R (ed): ''Nucelocytoplasmic Transport.'' Trendelenburg, Berlin: Springer-Verlag, pp 159–169, 1986.
11. Richardson WD, Roberts BL, Smith AL: Cell 44:77–85, 1986.
12. Welsh JD, Swimmer C, Cocke T, Shenk T: Mol Cell Biol 6:2207–2212, 1986.
13. Picard D, Yamamoto K: EMBO J 6:3333–3340, 1987.
14. Hall MN, Johnson AD: Science 237:1007–1012, 1987.
15. Silver PA, Brent R, Ptashne M: Mol Cell Biol 6:4763–4766, 1986.
16. Paucha E, Kalderon D, Richardson WD, Harvey RW, Smith AE: EMBO J 4:3235–3240, 1985.
17. Dingwall C, Sharnick SV, Laskey RA: Cell 30:449–458, 1982.
18. Newmeyer DD, Lucocq JM, Burglin TR, De Robertis EM: EMBO J 5:501–510, 1986.
19. Newmeyer DD, Finlay DR, Forbes DJ: J Cell Biol 103:2091–2102, 1986.
20. Feldherrr CM, Cohen RJ, Ogburn JA: J Cell Biol 96:1486–1490, 1983.
21. Goldfarb DS, Gariepy J, Schoolnik G, Kornberg RD: Nature 322:641–644, 1986.
22. Lanford RE, Kanda P, Kennedy RC: Cell 46:575–582, 1986.
23. Bonner WM: J Cell Biol 64:421–430, 1975.
24. Feldherr CM, Kallenbach E, Schultz N: J Cell Biol 99:2216–2222, 1984.
25. Finlay DR, Newmeyer DD, Price TM, Forbes DJ: J Cell Biol 104:189–200, 1987.
26. Astell AR, Ahlstrom-Jonasson L, Smith M, Tatchell K, Nasmyth KA, Hall BD: Cell 27:15–23, 1981.
27. Wallis JW, Hereford L, Grunstein M: Cell 22:799–805, 1980.
28. Smith MM, Andresson OS: J Mol Biol 169:663–690, 1983.
29. Choe J, Kolodrubetz D, Grunstein M: Proc Natl Acad Sci USA 79:1484–1487, 1982.
30. Adzuma K, Ogawa T, Ogawa H: Mol Cell Biol 4:2735–2744, 1984.
31. Roeder GS, Beard C, Smith M, Keranen S: Mol Cell Biol 5:1543–1553, 1985.

Molecular Biology of Intracellular Protein
Sorting and Organelle Assembly 193–202 (1988)

Biogenesis of Yeast Mating Pheromone a-Factor and *RAS* Proteins

Susan Michaelis and Scott Powers

Department of Biochemistry and Biophysics, University of California, San Francisco, California 94143 (S.M.); and Cold Spring Harbor Laboratory, Cold Spring Harbor, New York 11724 (S.P.)

The *Saccharomyces cerevisiae* pheromone **a**-factor is an extracellular signaling peptide involved in mating, whereas *RAS* proteins are localized to the plasma membrane and play a role in the modulation of cAMP levels in yeast. Although these proteins are targeted to different locations and perform unrelated functions, they appear to share a common step in biogenesis, since production of both active **a**-factor and active *RAS* proteins requires the *RAM* gene product. We review here the biochemical defects observed for *RAS* proteins in the *ram* mutant and also summarize our current views concerning the post-translational modification and localization of **a**-factor. We suggest that **a**-factor, like *RAS*, may contain a covalently linked fatty acid.

Key words: membrane localization, palmitylation, secretion, acylation, *Saccharomyces cerevisiae*, *RAM* gene

Intercellular signaling mediated by secreted peptide molecules such as hormones, growth factors, and neurotransmitters plays a role in a variety of developmental and physiological processes. The yeast *Saccharomyces cerevisiae* is a unicellular organism that provides a simple system for studying the mode of action and biogenesis of extracellular signaling molecules. This yeast exists in two alternative haploid cell types, **a** and α, which can mate to form a diploid zygote. Each cell type secretes a distinct peptide pheromone, **a**-factor or α-factor, respectively. These signaling molecules interact with receptors on cells of the opposite type to cause cell cycle arrest, induction of transcription of certain genes, and morphological alterations that are essential to the mating process [reviewed in 1].

STRUCTURE AND BIOGENESIS OF YEAST MATING PHEROMONES

α- Factor

The two mating pheromones differ substantially in their chemical properties and in the pathway of their maturation and secretion. Mature α-factor purified from the

Received June 18, 1987.

culture supernatant of α cells is 13 amino acids long. The majority of α-factor is derived from a precursor encoded by the *MFα1* gene, which contains an amino terminal signal sequence, sites for Asn-linked glycosylation, and four repeats of the mature peptide flanked by dibasic residues and separated by spacer regions [2]. A small amount of α-factor is derived from a similar precursor encoded by *MFα2*, which has only two repeats of the mature peptide [3,4]. Processing of the precursor is initiated by an endoproteolytic cut at the Lys-Arg dibasic residues mediated by the *KEX2* gene product, a calpain-like endoprotease [5–7]. Following this step, exoproteolytic amino and carboxy terminal trimming by a dipeptidyl aminopeptidase encoded by *STE13* [8] and a carboxypeptidase encoded by *KEX1* (H. Bussey, personal communication) yields the mature α-factor peptide. The α-factor precursor is targeted to the endoplasmic reticulum by its amino terminal signal sequence; subsequent glycosylation of the precursor and proteolytic processing occur during transit through the endoplasmic reticulum, Golgi, and secretory vesicles [9].

a-Factor

In contrast to α-factor, our understanding of the structure of **a**-factor and its pathway of biogenesis is far less complete. Chemical analysis of **a**-factor has been hampered by its extreme hydrophobicity and its tendency to aggregate with unknown components of the culture medium [10–12], making purification difficult. Nevertheless, Betz and Duntz [10,11] succeeded in purifying **a**-factor, based on its ability to cause morphological alterations and arrest the growth of α cells. They tentatively reported the amino acid sequence of two similar species of **a**-factor, 11 amino acids in length, which differ from one another at a single residue. Using this peptide sequence data, Brake et al. [13] constructed oligonucleotide probes and cloned two genes that contain sequences corresponding to the reported peptides. These genes, *MFa1* and *MFa2*, were shown to indeed encode **a**-factor, since, when present in cells on a high copy number plasmid, **a**-factor is overproduced [13; S. Michaelis and I. Herskowitz, unpublished observation]. Furthermore, we have shown that when these genes are deleted, the double *mfa1 mfa2* mutant produces no **a**-factor and is unable to mate. Each gene appears to contribute approximately equally to total **a**-factor activity, since single *mfa1* or *mfa2* mutants make half as much **a**-factor as a wild-type strain (S. Michaelis and I. Herskowitz, in preparation).

Examination of the predicted products encoded by *MFa1* and *MFa2* indicates that they are quite similar and reveals that mature **a**-factor must be processed from a precursor polypeptide, 36 amino acids long for *MFa1* and 38 amino acids long for *MFa2* [13] (Fig. 1). But serious questions remain concerning the exact structure of the mature pheromone, particularly at its carboxy terminal end. A synthetic 11-mer peptide containing the residues first proposed by Betz and Duntze to comprise mature **a**-factor (residues 22 to 32, Fig. 1) is inactive (J. Becker, personal communication and S. Michaelis, unpublished observation), as is a 15-mer peptide containing the additional four carboxy-terminal amino acids predicted by the DNA sequence (residues 22 to 36) [14]. Interestingly, however, a synthetic 12-mer terminating with Cys (residues 22 to 33) does exhibit low but detectable biological activity [14; J. Becker and F. Naider, personal communication]. Evidence for inclusion of Cys also comes from ^{35}S-sulfate labeling studies [7], which show that the mature extracellular species includes a sulfur-containing residue. Furthermore, in a recent report [15], Betz and coworkers have revised their initial findings and now suggest that **a**-factor is a 12-mer (Fig. 1) containing a

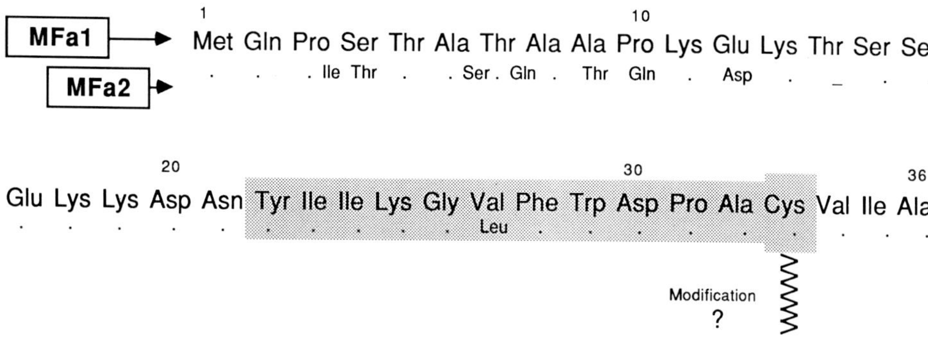

Fig. 1. Structure of **a**-factor precursor polypeptides. Amino acid sequence of the **a**-factor precursors as deduced from DNA sequence of the *MFa1* and *MFa2* [13] is shown. Corresponding residues of *MFa2* are aligned to those of *MFa1*. Dots indicate identical residues. Insertions and substitutions in *MFa2* are indicated. The dash signifies that the corresponding residue is absent in *MFa2*. Amino acids are numbered according to their position in the *MFa1* precursor. The amino acids present in mature **a**-factor (see text) are shaded. The jagged line represents the proposed fatty acid modification of Cys (see text).

carboxy-terminal Cys residue, which was previously overlooked due to an unsuspected modification.

Several lines of evidence raise the possibility that this carboxy terminal Cys residue contains a lipid modification: 1) the synthetic 12-mer exhibits only scant activity, suggesting it does not precisely correspond to mature **a**-factor and is thus missing a crucial component; 2) the low solubility and extreme hydrophobic nature of **a**-factor during purification suggest the presence of a nonpolar moiety, in addition to the identified amino acids; and 3) as will be discussed in detail below, the Cys-Val-Ile-Ala sequence (residues 33–36) at the carboxy terminus of the **a**-factor precursor shows significant similarity to the carboxy terminal sequence of yeast and mammalian *RAS* proteins, within which addition of the fatty acid palmitate to Cys is known to occur (16–18). The presence of a Cys residue with a nonpolar addition would not be detected using typical sequencing procedures and could account for the problems mentioned above in establishing the amino acid sequence of the **a**-factor carboxy terminus. Precedent for the existence of lipopeptide mating pheromones comes from studies of several heterobasidiomycetous yeasts [19–21]. Rhodotorucine A, an 11 residue long peptide secreted by *Rhodosporidium toruloides*, contains a carboxy terminal Cys residue modified by addition of a farnesyl moiety. Similarly, a pair of peptide pheromones from *Tremella mesenterica*, tremerogens A-10 (a 12-mer) and a-13 (a 13-mer), contain isoprenyl-modified Cys at their carboxy termini.

Another notable feature of the putative **a**-factor precursor is the absence of a hydrophobic amino terminal signal sequence. Furthermore, sites diagnostic for Asn-linked glycosylation are not present. The lack of these hallmark features for secreted proteins suggests that **a**-factor may not utilize the classical secretory pathway involving the endoplasmic reticulum, Golgi, and secretory vesicles, but instead may exit the cell via a novel route.

STE6, *STE14*, AND *STE16* GENE PRODUCTS ARE REQUIRED FOR a-FACTOR BIOGENESIS

We have employed a genetic approach to identify components involved in the processing and secretion of **a**-factor. Mutations causing sterility, i.e., the inability to mate, have been identified by a variety of selection and screening procedures. Among these are the *ste6* [22,23], *ste14* [24], and *ste16* [18,25] mutations, which confer sterility to **a** cells. **a** cells containing the *ste6*, *ste14*, or *ste16* mutations, while remaining responsive to α-factor, are unable to mate due to a defect in **a**-factor synthesis. α cells containing these lesions are mating-proficient and produce α-factor normally, indicating that the *ste6*, *ste14*, and *ste16* mutations interfere specifically with **a**-factor synthesis. Using Northern hybridization and gene-fusion analysis, we have demonstrated that these mutations do not affect transcription of the *MFa1* and *MFa2* genes (S. Michaelis and I. Herskowitz, in preparation). Thus the *STE6*, *STE14*, and *STE16* gene products are involved in post-transcriptional steps of **a**-factor biosynthesis and may participate in proteolytic processing, modification, or secretion.

Our current studies have focused on *STE16*. In addition to their **a** cell mating defect, *ste16* mutants exhibit a severe growth defect. Slow doubling times are observed at all temperatures, and growth is completely inhibited at very low temperatures (23°) [18; S. Michaelis, unpublished observation]. This slow growth behavior of *ste16* strains is not due to the absence of **a**-factor activity, since strains unable to produce **a**-factor for other reasons (such as deletion of the **a**-factor genes) grow normally. Furthermore, α haploids do not make **a**-factor but also exhibit slow growth when the *ste16* mutation is present. Thus other important cellular processes in addition to mating must be dependent on *STE16* function.

SUPH GENE PRODUCT IS REQUIRED FOR *RAS* FUNCTION

An important clue to understanding the nature of the *ste16* defect has come from studies of the yeast *RAS* genes, which encode proteins homologous to products of the mammalian *ras* oncogenes [26]. The yeast *RAS1* and *RAS2* proteins share common functional and structural properties with their mammalian counterparts. They bind GTP and GDP, exhibit GTPase activity, and are localized to the cytoplasmic face of the plasma membrane, although they lack a hydrophobic membrane spanning sequence [17,29–30]. Instead, they contain a covalently bound palmitate moiety that appears to be responsible, at least in part, for anchoring them to the membrane [16–18,30]. The function of *RAS1* and *RAS2* proteins in yeast is to modulate intracellular levels of cAMP [27,28].

A genetic selection, designed to identify proteins involved in *RAS* biosynthesis and function, yielded a class of mutants designated *supH*, which are defective in *RAS* activity [18]. The *supH* lesions do not reside within the *RAS1* or *RAS2* genes, but instead in a gene that influences *ras* activity. *supH* mutants were shown to be severely attenuated in *RAS*-dependent adenylate cyclase activity and are temperature-sensitive for growth. In addition, they exhibited an unexpected phenotype: **a** cell sterility due to an inability to produce functional **a**-factor activity. This finding was surprising, since there was no previously known connection between *RAS* and **a**-factor; it prompted our two groups (S. Michaelis and I. Herskowitz, studying mating and **a**-factor biogenesis, and S. Powers and M. Wigler, investigating *RAS*) to carry out the collaborative study [18] described below.

STE16 AND *SUPH* ARE THE SAME GENE, REDESIGNATED *RAM* (FOR *RAS* AND a-FACTOR MATURATION)

We were struck by the similarity between the *ste16* and *supH* mutants. While isolated independently, based on the unrelated phenotypes of sterility and altered *RAS* activity, they exhibited the similar properties of: 1) defective growth (at high temperature in the case of *supH* and all temperatures for *ste16*); and 2) absence of **a**-factor activity. These observations led us to consider that *SUPH* and *STE16* might be the same gene. Complementation and alellism tests proved this to be the case. Furthermore, both of our groups had cloned *STE16* and *SUPH* by different means, and comparison of our plasmids revealed that they contained a common segment of DNA. The cloned *STE16/SUPH* gene is being used to construct a null allele that is expected to cause a lethal phenotype in haploids, based on the temperature-sensitive and cold-sensitive phenotypes of the *supH* and *ste16* mutants described above. We have redesignated the *STE16/SUPH* gene as *RAM* (for *RAS* and **a**-factor maturation) [18].

Is there some similar feature shared by *RAS* and **a**-factor that could explain their common requirement for *RAM* function? At the outset of our studies, we noticed an intriguing amino acid sequence similarity between these two proteins. Certain eukaryotic proteins, including all known *RAS* proteins, share the common carboxy terminal consensus sequence Cys-A-A-X_{COOH}, (where A is an aliphatic amino acid and X is the carboxy terminal amino acid) (Table I). In *RAS* proteins, palmitic acid is attached, presumably via a thioester bond, to the Cys residue in this conserved region [16]. A mutationally altered mammalian *H-ras* protein with Ser substituted for Cys does not become acylated, fails to localize to the membrane, and fails to transform cells, indicating

TABLE I. Common C-Terminal Sequences in *RAS* Proteins, a-Factor, and Other Proteins*

Gene	Species	C-terminal sequence						Reference
RAS1	*S. cerevisiae*	Gly	Cys	*Cys*	Ile	Ile	Cys	26
RAS2	*S. cerevisiae*	Gly	Cys	*Cys*	Ile	Ile	Ser	26
ras1	*S. pombe*	Cys	Lys	*Cys*	Val	Leu	Ser	41
Dd-ras	*Dictyostelium*	Lys	Gln	*Cys*	Leu	Ile	Leu	42
Dras3	*Drosophila*	Val	Pro	*Cys*	Val	Leu	Leu	43
H-ras	Human	Cys	Lys	*Cys*	Val	Leu	Ser	44
K-ras	Human	Lys	Lys	*Cys*	Ile	Ile	Met	44
N-ras	Human	Leu	Pro	*Cys*	Val	Val	Met	44
Transducin-γ	Bovine	Gly	Gly	*Cys*	Val	Ile	Ser	45
rho	*Aplysia*	Gly	Gly	*Cys*	Val	Val	Leu	46
RHO1	*S. cerevisiae*	Lys	Lys	*Cys*	Val	Leu	Leu	37
RHO2	*S. cerevisiae*	Asn	Cys	*Cys*	Ile	Ile	Leu	37
MFa1	*S. cerevisiae*	Pro	Ala	*Cys*	Val	Ile	Ala	13
MFa2	*S. cerevisiae*	Pro	Ala	*Cys*	Val	Ile	Ala	13
Consensus[a]				CYS	A	A	X_{COOH}	

*Reprinted from reference 18 with permission from the publisher, Cell Press.

[a]In the consensus sequence Cys-A-A-X, A is an aliphatic amino acid (either valine, leucine, or isoleucine), and X is the C-terminal amino acid. Transducin-γ represents the γ-subunit of bovine retinal transducin.

the importance of the Cys residue and suggesting that acylation is a critical event for membrane targeting and function [31]. Like *RAS*, the carboxy terminus of the predicted **a**-factor precursor also contains the Cys-A-A-X_{COOH} consensus sequence (Fig. 1 and Table 1), prompting us to suggest that **a**-factor may be an acylated protein and that *RAM* activity is required for acylation of both **a**-factor and *RAS*.

ROLE OF THE *RAM* GENE PRODUCT

Acylation, Membrane Localization, and Maturation of *RAS* Proteins are Defective in *RAM* Mutants

In both mammalian and yeast cells, *RAS* proteins are initially synthesized as soluble precursors that mature into membrane-bound products. Their biogenesis involves three observable changes: 1) addition of fatty acid; 2) localization to the membrane; and 3) a shift in mobility observed by SDS-PAGE analysis [17,18,29–32]. This latter change to an apparently lower molecular weight species (by ~ 1 kD) likely reflects a processing or modification event that has not yet been identified and does not result from addition of fatty acid, which appears to have no effect on mobility. We will refer to this step as "maturation." Tamanoi [33] has shown that no amino terminal proteolytic processing occurs for *RAS*; carboxy terminal proteolytic processing remains to be investigated. It has not yet been established whether the changes observed during *RAS* biogenesis occur in sequential fashion or whether they occur coordinately. In our studies these processes are not temporally separable.

We asked whether any of these processes are altered in the *ram* mutant. We performed pulse-chase labeling, following by cell fractionation in wild-type and mutant strains and analyzed the immunoprecipitated RAS proteins by SDS-PAGE. In the wild-type strains, *RAS* is initially found as a slow-migrating, soluble species that matures to a faster-migrating, membrane-bound form with a relatively long half-time (approximately 20 minutes, using standard growth conditions). In contrast, neither maturation nor membrane localization occur in the *ram* mutant. Instead, the precursor simply accumulates in the cytoplasm. In further experiments, we also examined acylation. Cells were labeled with ^{3}H-palmitate and analyzed by SDS-PAGE after immunoprecipitation. While *RAS* becomes ^{3}H-labeled in the wild type strain, no detectable labeling of *RAS* occurs in the mutant. Together, these results demonstrate that the *RAM* product is required for membrane targeting, maturation, and acylation of *RAS* proteins in yeast and support the notion that acylation promotes membrane localization, which, in turn, permits functioning of RAS.

One attractive hypothesis is that *RAM* encodes the acyltransferase responsible for adding palmitate to *RAS* proteins. Biochemical identification of the presumed palmityl-transferase and determination of its presence or absence in *ram* mutants will allow a direct test for this proposal. Recently, an enzyme specific for addition of myristate to the amino terminal residue of particular peptides has been purified from yeast [34–36]. An analogous strategy, employing peptides containing the sequence Cys-A-A-X_{COOH}, may by useful for identifying the palmityl-transferase implicated in *RAS* modification. An alternative hypothesis for the function of the *RAM* product is that it acts to "prepare" *RAS* for acylation, perhaps by promoting the maturation step. Were this the case, maturation must occur prior to addition of palmitate to explain why *both* maturation and acylation are absent in *ram* mutants. Although observations consistent with this sequence of events have been reported [17,32], clear-cut evidence supporting such a pathway is lacking. Determination of the precise biochemical role of the *RAM* gene product should clarify the pathway of *RAS* biogenesis.

Subsequent to our isolation of *ram* mutants, Tamanoi and coworkers [33] characterized a mutant that they designate *dpr*. We have shown by an allelism test that *DPR* is the same gene as *RAM* (S. Michaelis, unpublished results). Biochemical analysis of *RAS* carried out by this group agrees in large measure with ours, namely, they found that maturation, membrane localization, and acylation are all impaired in the *dpr* mutant. However, they observed a residual amount of acylation of RAS. While we feel it is most likely that this simply represents leakiness of the *dpr* allele of *RAM*, they favor the hypothesis that acyltransferase activity remains intact and, instead, that the primary defect in *ram* mutants is in the maturation step.

Other Proteins Also Require *RAM* Function

In addition to *RAS*, we have shown that acylation of several other yeast membrane proteins is impaired in *ram* mutants [18]. By SDS-PAGE analysis of a total membrane fraction from ^{3}H-palmitate-labeled wild-type cells, we observe five major labeled proteins of approximate molecular weights 140, 90, 60, 30, and 25 kD. The three largest of these fail to become acylated in a *ram* mutant. While their identity has not been established, it is tempting to speculate that some of these proteins may share with *RAS* and **a**-factor the Cys-A-A-X_{COOH} sequence. Other yeast proteins that terminate with this sequence are the *RAS*-homologous *RHO1* and *RHO2* gene products [37] (Fig. 1), although there is as yet no evidence for their acylation, nor are they likely to be among the membrane proteins observed in the above experiment, based on their predicted molecular weight (21 kD).

Our finding that there are two classes of ^{3}H-palmitate-labeled proteins distinguished by their pattern of acylation in the *ram* mutant may indicate the existence of two separate pathways for addition of this fatty acid to proteins. Certain viral glycoproteins and other eukaryotic membrane proteins that traverse the secretory pathway are thought to be palmitylated in the Golgi [38]. *RAS*, by contrast, does not appear to utilize the secretory pathway and instead acquires fatty acid in the cytoplasm or at the plasma membrane. It is thus tempting to speculate that there are two differently localized acyltransferase enzymes, only one of which remains operative in the *ram* mutant.

APPROACHES FOR DETERMINING THE PATHWAY OF a-FACTOR BIOGENESIS

Our observation of the Cys-A-A-X_{COOH} sequence in the **a**-factor precursor, in conjunction with analysis of the *RAM* mutant, has led us to propose that **a**-factor is a lipopeptide. Our current working model is that **a**-factor is an oligopeptide 12 amino acids long, which terminates with a palmityl-Cys residue (Fig. 1). According to this view, **a**-factor biogenesis must involve: 1) carboxy and amino terminal protolytic processing of the precursor; 2) acylation; and 3) a specialized export apparatus, since typical secretion signals are absent. The *STE6*, *STE14*, and *RAM* gene products are likely candidates for mediating some or all of these steps. A generalized scheme for **a**-factor biogenesis is shown in Figure 2. Our depiction is purposely vague in suggesting an order for these steps and in assigning roles to *STE6* and *STE14*, reflecting the rudimentary state of our knowledge. The primary role of these gene products could be in proteolytic processing or, alternatively, in promoting export. Both genes have been cloned, and

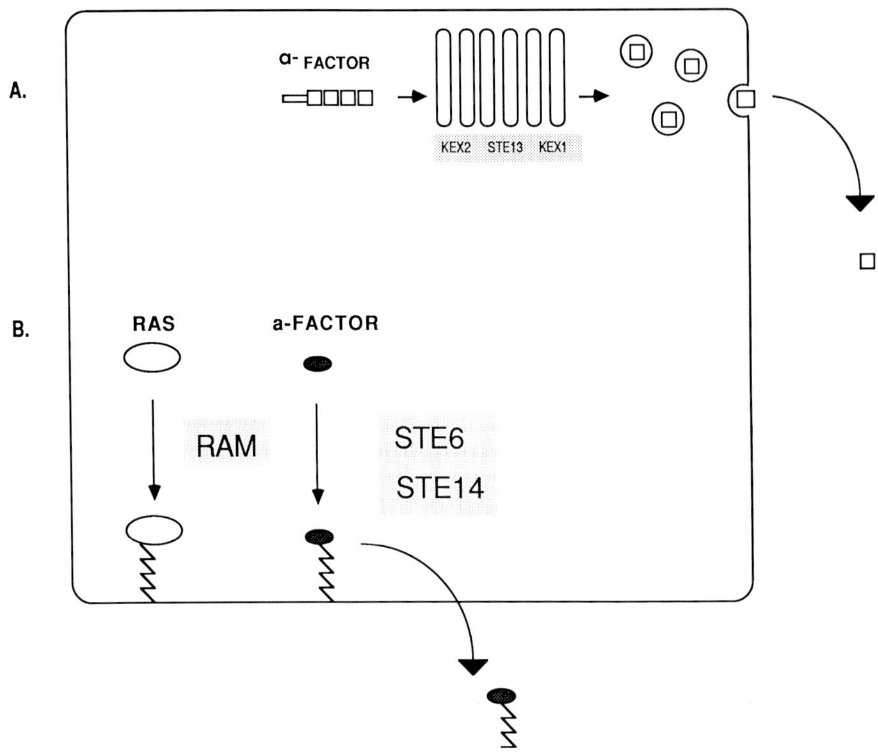

Fig. 2. Localization of α-factor, **a**-factor, and *RAS*. **A)** the α-factor precursor (joined squares) is targeted to the secretory pathway for export. (Clustered ovals indicate the endoplasmic reticulum and Golgi; circles represent secretory vesicles.) Within the secretory apparatus, the *KEX2*, *STE13*, and *KEX1* gene products mediate proteolytic processing events that yield mature, extracellular α-factor (single square). **B)** Localization of *RAS* (unfilled oval) to the plasma membrane does not involve the secretory pathway. Instead, the cytoplasmic *RAS* precursor undergoes maturation and addition of palmitate (jagged line), which promotes targeting of mature *RAS* protein to the inner face of the plasma membrane. Acylation (or maturation) are mediated by the *RAM* gene product (see text). Biogenesis of **a**-factor (filled oval), like *RAS*, requires *RAM*. In the model shown here, the first step in **a**-factor secretion is *RAM*-dependent acylation of the **a**-factor precursor, resulting in localization to the membrane. A subsequent event is then proposed to promote secretion of **a**-factor through the plasma membrane to its extracellular location. The *STE6* and *STE14* gene products may mediate proteolytic processing of the **a**-factor precursor or its secretion. The order of steps promoted by *STE6*, *STE14*, and *RAM* has not yet been determined.

null alleles have been constructed for both [22; S. Michaelis and K. Wilson, unpublished results]. Determination of the nucleotide sequence of these genes and biochemical analysis of the mutants should provide more insight into their function.

Although the cytoplasmic membrane is the final destination for *RAS*, **a**-factor must eventually achieve an extracellular location. Perhaps by undergoing *RAM*-mediated membrane targeting via acylation, **a**-factor accomplishes the first step of its transport out of the cytoplasm. Subsequently, a specialized carrier, pore, or flipping mechanism might promote its exit through the plasma membrane. Although not mechanistically understood, such a novel export apparatus has been implicated in the secretion of several bacterial proteins, notably an *Escherichia coli* haemolysin [39] and microcin [40].

Clearly, antibodies recognizing mature and precursor forms of **a**-factor will be useful for identifying the intermediates in **a**-factor biosynthesis and their cellular location. Several groups, including our own, have recently obtaned antibodies against synthetic peptides containing **a**-factor sequences. Making use of such antibodies, Sterne and Thorner [7] have preliminarily reported the accumulation of cell-associated precursor forms of **a**-factor in *ste6* and *ste14* mutants. The question of acylation can also be addressed using antibodies. Is **a**-factor precursor labeled with ^{3}H-palmitate? If so, is addition of fatty acid an early or late step, and is it blocked in *ram* mutants?

CONCLUSIONS

Although the **a**-factor and α-factor pheromones are both secreted from the cell, they appear to use very different strategies to achieve extracellular localization. α-factor employs the classical secretory pathway, while **a**-factor apparently utilizes a novel pathway for export. An insight into this novel pathway has come from identification of the *RAM* gene, described here. These studies have revealed a surprising link in the biogenesis of **a**-factor and *RAS* proteins. Characterization of the *ram* mutant has suggested that the export of **a**-factor, like the targeting of *RAS* to the plasma membrane, may involve a fatty acid modification. Further genetic and biochemical analysis of **a**-factor and *RAS* biogenesis will likely provide new insight into the role of fatty acid acylation in protein localization.

ACKNOWLEDGMENTS

We thank Nancy Craig, Lorraine Marsh, and Paul Sternberg for helpful comments and critical review of the manuscript. Portions of the work described here were supported by an NIH fellowship (F32 GM09934) to S.M. and a Leukemia Society of America fellowship to S.P.

REFERENCES

1. Sprague GF, Blair LC, Thorner J: Annu Rev Microbiol 37:623, 1983.
2. Kurjan, J, Herskowitz I: Cell 30:933, 1982.
3. Kurjan J: Mol Cell Biol 5:787, 1985.
4. Singh A, Chen EY, Lugovoy JM, Chang CN, Hitzman RA, Seeburg PH: Nucleic Acids Res 11:4049, 1983.
5. Fuller RS, Brake AJ, Julius DJ, Thorner J: In Gething M-J (ed): "Protein Transport and Secretion." Cold Spring Harbor: Cold Spring Harbor Press, 1985, pp 97–102.
6. Julius D, Brake A, Blair L, Kunisawa R, Thorner J: Cell 37:1075, 1984.
7. Fuller R, Brake A, Sterne R, Kunisawa R, Barnes D, Flessel M, Thorner J: In Hicks J (ed): "Yeast Cell Biology." New York: Alan R. Liss, 1986, pp 461–476.
8. Julius D, Blair L, Brake A, Sprague G, Thorner J: Cell 32:839, 1983.
9. Julius D, Schekman R, Thorner J: Cell 36:309, 1984.
10. Betz R, Duntze W: Eur J Biochem 95:469, 1979.
11. Betz R, Manney TR, Duntze W: Gamete Res 4:571, 1981.
12. Strazdis JR, MacKay VL: J Bacteriol 151:1153, 1982.
13. Brake AJ, Brenner C, Najarian R, Laybourn P, Merryweather J: In Gething M-J (ed) "Protein Transport and Secretion." Cold Spring Harbor: Cold Spring Harbor Press, 1985, pp 103–108.
14. Sterne R, Thorner J: Yeast Genetics and Molecular Biology Meeting Abstracts, 1986, p 10.
15. Betz R, Crabb JW, Meyer HE, Wittig R, Duntze W: J Biol Chem 262:546, 1987.
16. Buss JE, Sefton BM: Mol Cell Biol 6:116, 1986.

17. Fujiyama A, Tamanoi F: Proc Natl Acad Sci USA 83:1266, 1986.
18. Powers S, Michaelis S, Broek D, Santa Anna-A S, Field J, Herskowitz I, Wigler M: Cell 47:413, 1986.
19. Kamiya Y, Sakurai A, Tamura S, Takahashi N, Tsuchiya E, Abe K, Fukui S: Agric Biol Chem 43:363, 1979.
20. Sakagami Y, Isogai A, Suzuki A, Kitada C, Fujino M: Agric Biol Chem 43:2643, 1979.
21. Sakagami Y, Yoshida M, Isogai A, Suzuki A: Agric Biol Chem 45:1045, 1981.
22. Wilson KL, Herskowitz I: Mol Cell Biol 4:2420, 1984.
23. Rine J: "Regulation of transposition of cryptic mating genes in *Saccharomyces cerevisiae*." PhD dissertation: University of Oregon, Eugene, 1979.
24. Blair LC: "Genetic analysis of mating type switching in yeast." PhD dissertation: University of Oregon, Eugene, 1979.
25. Wilson KL, Herskowitz I: Genetics 155:441, 1987.
26. Powers S, Kataoka T, Fasano O, Goldfarb M, Strathern J, Broach J, Wigler M: Cell 36:607, 1984.
27. Broek D, Samiy N, Fasano O, Fujiyama A, Tamanoi F, Northrup J, Wigler M: Cell 41:763, 1985.
28. Toda T, Uno I, Ishikawa T, Powers S, Kataoka T, Broek D, Cameron S, Broach J, Matsumoto K, Wigler M: Cell 40:27, 1985.
29. Clark SG, McGrath JP, Levinson AD: Mol Cell Biol 5:2746, 1985.
30. Shih TY, Weeks MO, Gross P, Dhar R, Oroszlan S, Scolnick EM: J Virol 42:253, 1982.
31. Willumsen BM, Norris K, Papageorge AG, Hubbert NL, Lowry DR: EMBO J 3:2581, 1984.
32. Fujiyama A, Samiy N, Rao M, Tamanoi F: In Hicks J (ed): "Yeast Cell Biology." New York: Alan R. Liss, 1986, pp 125–149.
33. Fujiyama A, Matsumoto K, Tamanoi F: EMBO J 6:223, 1987.
34. Towler D, Glaser L: Proc Natl Acad Sci USA 83:2812, 1986.
35. Towler DA, Adams SP, Eubanks SR, Towery DS, Jackson-Machelski E, Glaser L, Gordon JI: Proc Natl Acad Sci USA 84:2708, 1987.
36. Towler DA, Eubanks SR, Towery DS, Adams SP, Glaser L: J Biol Chem 262:1030, 1987.
37. Madaule P, Axel R, Meyers A: Proc Natl Acad Sci USA 84:779, 1987.
38. Schmidt MFG: Curr Top Microbiol Immunol 102:101, 1983.
39. Wagner W, Vogel M, Goebel W: J Bacteriol 154:200, 1983.
40. Daragnino J, Herrero M, Furlong D, Moreno F, Kolter R: Proteins 1:230, 1986.
41. Fukui Y, Kaziro Y: EMBO J 4:687, 1985.
42. Reymond CD, Gomer RH, Mehdy MC, Firtel RA: Cell 39:141, 1984.
43. Schejter ED, Shilo B-Z: EMBO J 4:407, 1985.
44. Taparowsky E, Shimizu K, Goldfarb M, Wigler M: Cell 34:581, 1983.
45. Hurley JB, Fong HKW, Teplow DB, Dreyer WJ, Simon MI: Proc Natl Acad Sci USA 81:6948, 1984.
46. Madaule P, Axel R: Cell 41:31, 1985.

Molecular Biology of Intracellular Protein
Sorting and Organelle Assembly 203–213 (1988)

Requirements for Post-Translational Movement of Proteins Through the Endoplasmic Reticulum Membrane of Yeast and Mammalian Cells

William J. Hansen and Peter Walter

Department of Biochemistry and Biophysics, University of California, San Francisco, California 94143

Protein translocation across the mammalian endoplasmic reticulum (ER) membrane is generally a cotranslational event in which the nascent polypeptide chain is moved across the lipid bilayer as it emerges from the ribosome. It recently became apparent that, in principle, eukaryotic endoplasmic reticulum membranes are also capable of accepting certain proteins as translocation substrates after they have been completely synthesized. Because translocation can occur post-translationally, the distinction between the co- and post-translational modes becomes a kinetic argument, where the degree of coupling observed between translation and translocation is a function of the relative rates of the two respective processes. While the fundamental mechanism of membrane translocation may be the same co- and post-translationally (i.e., proteins are likely to use the same translocation machinery), the molecular requirements of the targeting reaction and molecular requirements on the structure of the substrate protein per se appear to differ.

In particular, we find that cotranslational targeting involves the function of the signal recognition particle (SRP) and the SRP receptor. These components ensure that ribosomes synthesizing a nascent protein are specifically brought to the correct intracellular membrane system. Once targeted, as yet unidentified proteins are involved in the formation of a functional ribosome-membrane junction and the translocation of the growing protein chain across the membrane. Most interestingly, work from Reid Gilmore's laboratory has recently demonstrated that the formation of the ribosome membrane junction requires a GTP binding protein, but no hydrolysis of the nucleoside triphosphate.

In contrast, post-translational translocation across ER membranes requires ATP, and nonhydrolyzable analogs will not substitute. It is our current working hypothesis that this energy requirement reflects the need to unfold either completely or partially the translocation substrate to render it translocation-competent. Not every protein can utilize the post-translational mode of transmembrane movement, presumably because of tertiary structural features that may not be compatible with translocation. In addition, folding may be too tight to be reversible, even upon

Received June 29,1987.

action of an energy-consuming unfoldase. For example, we find for yeast invertase that the full-length molecule is incompetent for translocation if polypeptide elongation is inhibited, while a form that is perturbed in its folding behavior (a truncated form missing about half of the invertase molecule from the carboxy terminal end) is competent. According to this hypothesis, the cotranslational targeting mode could have evolved to overcome restrictions posed by tertiary structural features in the substrate proteins. The proteins are targeted before their synthesis is completed and thus before they have a chance to fold into structures that then may be incompatible with the (common) translocation process.

Key words: endoplasmic reticulum, protein secretion, *Saccharomyces cerevisiae*, yeast, post-translational protein translocation

Most of what we know about the mechanism of protein translocation into the endoplasmic reticulum (ER) of mammalian cells has resulted from the ability to reconstitute the process in in vitro reactions [1]. Essential components catalyzing the translocation of proteins into the ER of canine pancreas were purified based upon in vitro assays devised to measure their functions in targeting presecretory proteins to microsomal vesicles. The signal recognition particle (SRP, which consists of six polypeptides and a molecule of RNA) was purified based upon its ability to promote the cotranslational translocation of presecretory proteins synthesized in a wheat germ extract (which contains insufficient endogenous SRP) across SRP-depleted microsomal vesicles [2]. A membrane receptor, the SRP receptor [3], or docking protein [4] was purified based upon its ability to interact with SRP in a reaction leading to the targeting and translocation of the presecretory protein [for review, see 5,6].

The functional domains of SRP were dissected into a domain that was required to slow or halt the elongation of the presecretory polypeptide until the other domain required for signal sequence binding and translocation-promoting activity could effect an interaction of the nascent polypeptide with the ER membrane [7,8]. An SRP lacking the elongation arrest domain has been created by reconstitution and employed to show that the elongation arrest is not an essential step in the translocation process [7,8].

Recently it has been demonstrated that translocation into the ER in vitro can be initiated late during the translation of precursor proteins [9,10]. These findings are consistent with the lack of an absolute requirement for elongation arrest prior to the initiation of translocation. Nevertheless, there appears to be a window of time during elongation beyond which translocation can not be initiated [7,9,11,12]. These studies indicated that in some cases large portions of a ribosome-associated nascent protein can remain translocation-competent after synthesis.

Mathematical models of the translation and translocation processes have recently been formulated and combined to support a hypothesis regarding the degree of coupling of translocation to translation. According to this hypothesis, the window of translocation competence of a protein during polypeptide elongation may vary according to the affinity of each particular signal peptide for SRP and whether or not previously or subsequently translated protein sequence can render the signal peptide inaccessible [13]. The model is consistent with the notion that in some cases large segments of secretory proteins may be translocated across eukaryotic microsomes in a manner not tightly coupled to protein synthesis, and even that some proteins may still be translocated post-translationally, as they have been observed to be across the *E. coli* plasma membrane [for reviews, see 14,15] and across the mitochondrial [16,17] and chloroplast envelopes [18].

TRANSLOCATION OF PROTEINS INTO THE ENDOPLASMIC RETICULUM DOES NOT ALWAYS REQUIRE CONCURRENT PROTEIN SYNTHESIS

During the past year, several groups have reported upon data that has shown that secretory and membrane protein translocation into the endoplasmic reticulum of yeast and mammalian cells can indeed occur in vitro by a mechanism that does not require on-going protein synthesis [20–25,32]. We present here a comparison of the data recently obtained in in vitro reactions reconstituting protein translocation across yeast and mammalian microsomes. We first examine specificity with respect to the protein substrate and the components of the translocation machinery that participate in the post-translational translocation process. Second, we discuss the energy requirements of the translocation process.

BEHAVIOR OF DIFFERENT PROTEIN SUBSTRATES AND FACTORS REQUIRED

Post-Translational Translocation Across Yeast Microsomal Membranes

Prepro-α-factor. Surprisingly, the first substrate employed in the reconstitution of secretory protein translocation across yeast microsomal vesicles, prepro-α-factor, was found to be translocated as efficiently post-translationally as it was cotranslationally [20,21,23]. Addition of sufficient cycloheximide to block polypeptide chain elongation did not diminish translocation when membranes were subsequently added. Furthermore, in contrast to all the proteins studied using mammalian microsomal membranes, except for a small artificial fusion protein (consisting of the signal sequence of β-lactamase fused to a fragment of globin) [22], the bacteriophage M13 pro-coat protein [26] and prepro-melittin [25] translocation was still observed after removal of the ribosomes from the reaction mixture prior to membrane addition [20,23]. Therefore this reaction is truly post-translational, since synthesis has been terminated and the newly synthesized protein has been released from the ribosome.

There is evidence that membrane and soluble proteins are required during the post-translational translocation of prepro-α-factor. Pretreatment of the microsomal vesicles with the alkylating reagent *N*-ethylmaleimide (which reacts with the sulphydryl group on cysteine residues) or with high concentrations of trypsin blocked the post-translational translocation of prepro-α-factor [20]. Inhibition of translocation by treatment of the membrane fraction with alkylating reagents or with proteases indicates that the function of proteins associated with the yeast microsomal vesicles are required during the translocation reaction. Whether the sensitivity to *N*-ethylmaleimide was due to the inactivation of the yeast analog of the mammalian SRP receptor [which is known to be inactivated by this reagent (3)], or whether it represents inactivation of some other protein component of the translocation machinery, is not certain. Others have observed that *N*-ethylmaleimide inactivates a component or components, distinct from the SRP receptor, which is (are) required during translocation into mammalian microsomal vesicles [25,27]. Two observations make it unlikely that a yeast SRP receptor analog is required during the post-translational translocation of prepro-α-factor. First, the translocation of prepro-α-factor can be ribosome-independent [20,23], while translocation of proteins that require SRP and SRP receptor functions also require that the precursor protein be associated with the ribosome. Second, the concentration of protease that was required to render yeast microsomes translocation-incompetent was 100 times higher than that required to inactivate the SRP receptor of mammalian microsomes [3].

Most recently, an assay for a soluble factor that stimulates the post-translational translocation across yeast microsomal vesicles of prepro-α-factor synthesized in the wheat germ extract has been described [28]. This factor does not appear to contain an RNA, as does the mammalian SRP [28]. Whether this factor stimulates the cotranslational translocation process or post-translational translocation of the precursor synthesized in the yeast extract was not determined. It is likely that a soluble factor is required during translocation of prepro-α-factor (which contains no cysteine residues), synthesized in the yeast translation system, because *N*-ethylmaleimide treatment of the soluble fraction inhibits translocation (our unpublished observations).

By asking whether prepro-α-factor was unique in its ability to be translocated post-translationally across yeast microsomal vesicles, we could determine whether the mechanism of transmembrane movement of proteins could in general be independent of polypeptide elongation. Toward this end, we chose to examine the behavior of two other yeast glycoprotein precursors, prepro-carboxypeptidase Y and pre-invertase.

Prepro-carboxypeptidase Y. The 59-kD precursor to this vacuolar glycoprotein has been shown in vivo to utilize the early stages of the yeast secretory pathway. During its biogenesis, carboxypeptidase Y is modified by addition of asparagine-linked core oligosaccharides upon entry into the ER [29]. We have used the addition of core oligosaccharides and protection from digestion by added protease afforded by the microsomal membrane after translocation as criteria to study the transmembrane movement of the carboxypeptidase precursor. We have found that prepro-carboxypeptidase Y can also be translocated in an elongation-independent reaction, although this process is not as efficient as that found for the α-factor precursor [47]. The mechanism by which a protein traverses the microsomal membrane post-translationally has not been defined. In spite of its molecular size, it must be in such a conformation, or be able to adopt a conformation [possibly with the help of accessory proteins, as suggested for prepro-α-factor synthesized in the wheat germ extract (see above)] that is competent to be targeted via its signal sequence and to move through the translocation apparatus.

Pre-invertase. The full-length pre-invertase molecule does not appear to remain competent to be translocated independent of chain elongation, although it is nearly identical in molecular weight to prepro-carboxypeptidase Y. However, it can be translocated cotranslationally in vitro [47]. Thus, unlike prepro-α-factor or prepro-carboxypeptidase Y, it must be unable to adopt a conformation suited for translocation after its synthesis has stopped. Invertase is known to form dimers in the ER and in the cytoplasm of yeast cells. Perhaps accessory proteins contributing to the post-translational translocation process cannot act upon proteins that form very tightly folded structures or readily form oligomers. The former constraint has been clearly demonstrated to apply to the import of a fusion protein into yeast mitochondria [30]. In addition, evidence has recently been obtained from a study of protein translocation in *E. coli* cells suggesting that a kinetic competition exists between protein folding and translocation of the maltose binding protein [31].

Contrary to the findings for the full-length invertase precursor, a form of invertase that is perturbed in its folding behavior (a truncated form missing about one half of the invertase molecule from the carboxy terminal end) can be translocated independent of protein elongation, albeit at a lower efficiency than has been observed for prepro-α-factor [47]. Elongation of the protein beyond this point to its full size may either render the signal sequence inaccessible and thus prevent proper targeting or cause the protein to fold into a conformation that cannot be easily unfolded prior to or during translocation.

It has been reported previously that the truncated invertase molecule (approximately one half the size of the full-length molecule), which has been released from the ribosome by termination, can not be translocated post-translationally [21]. We have investigated the elongation-independent translocation of this truncated invertase encoded by an mRNA with or without a termination codon. We have found that both forms can be translocated independent of polypeptide elongation into yeast microsomes.

Given that the post-translational translocation across yeast microsomes requires membrane and soluble proteins [20,28] and a source of ATP (see below), the efficiency of the post-translational translocation of particular protein will very likely be a function of the relative need of the protein for the participation of the soluble factor(s) and ATP-requiring component. Therefore the concentration of the soluble factors, ATP, and microsomal membranes present in experiments carried out by different investigators will probably play a role in whether post-translational translocation of a particular protein will be detected.

We have found that translocation across yeast microsomal vesicles that is not coupled to polypeptide elongation translocation is not limited to prepro-α-factor. Therefore the mechanism of transmembrane movement of proteins across the yeast ER may in general include steps that do not occur tightly coupled to protein elongation. However, there are clearly examples of proteins (e.g., full-length pre-invertase) that must initiate translocation before polypeptide elongation has proceeded too far. Protein substrates in this latter class should prove useful in the search for yeast proteins required for the efficient coupling of translocation and translation.

Post-Translational Translocation Across Mammalian Microsomal Membranes.

The human glucose transporter. This protein, which spans the membrane as many as 12 times [19], was observed to insert into pancreatic microsomes after it was completely synthesized in the reticulocyte lysate [32], although the efficiency of the process was reported to be about 20-fold lower than the cotranslational process. Truncation of the coding sequence, such that the amino terminal two thirds of the transporter was synthesized without being released from the ribosome, resulted in an enhancement of the translocation efficiency. The resulting increase in efficiency was postulated to be due to a decreased intermolecular aggregation of the truncated versus the full-length transporter.

Elongation-independent insertion and glycosylation of the truncated glucose transporter was sensitive to RNAse A, suggesting, albeit rather indirectly, that SRP may be required for these processes. Further support for the notion that SRP was required during the ribosome-dependent but elongation-independent translocation process came from an experiment in which micrococcal nuclease digestion of the extract containing the precursor was followed by addition of microsomal membranes. In this experiment only unglycosylated precursor was bound to the membranes. Addition of purified SRP resulted in glycosylation of some of the membrane-bound material, which supports the idea that proper insertion of the transporter (resulting in oligosaccharide addition) required SRP. It would appear consistent with the data obtained that the difference in the efficiencies of translocation of full-length glucose transporter and the truncated molecule may result from the ability of SRP to catalyze the translocation of the transporter precursor protein only in the latter case, when the polypeptide remains attached to the ribosome.

Gene fusions to globin. Two gene-fusion proteins containing globin at their carboxyl termini, one with the signal sequence of the prokaryotic secretory protein β-lactamase at its amino terminus and the other with an amino terminal fragment of the membrane protein bovine opsin, were shown to be translocated in the absence of concurrent protein synthesis [22]. In the case of the β-lactamase globin fusion, however, post-translational translocation was not detected after the full-length fusion was synthesized. Translocation was only detected if membranes were added after polypeptide elongation was inhibited when a truncated fusion protein was synthesized from a mRNA lacking a termination codon, or if the globin coding sequence was shortened to only 78 amino acids by insertion of a termination codon early in the coding region of the mRNA. The role of the association of the former substrate with the ribosome was postulated to be one in which the interaction results in an increased translocation competence of the polypeptide, possibly through the prevention of the folding of a structure deleterious to targeting or translocation. This association with the ribosome should also affect whether SRP is able to catalyze the translocation process. An observation consistent with this suggestion is that the elongation-independent translocation of the truncated lactamase-globin fusion that has not been released from the ribosomes of the SRP-deficient wheat germ extract does, in fact, require addition of exogenous SRP (R. Rothman and V. Lingappa, unpublished observations). The coding sequence following the signal sequence of the fusion protein containing only 78 amino acids of globin must not interfere with translocation, since post-translational translocation could be detected after release of this protein from the ribosome.

Human preplacental lactogen. Translocation of a fully synthesized mammalian secretory protein, the human preplacental lactogen, which is synthesized in the reticulocyte lysate, has been reported to occur in an elongation independent fashion [24]. However, *only* the ribosome-associated precursor molecules were observed to be translocated. This report showed that a bona fide mammalian secretory protein can traverse the mammalian ER membrane in a reaction that is independent of polypeptide elongation.

Honey bee prepro-melittin. The glucose transporter proteins were not translocated independent of elongation if they were synthesized in a *wheat germ* translation system (which contains no endogenous SRP) [32]. In contrast, the insect protein prepro-melittin synthesized in a wheat germ extract could be translocated post-translationally at an efficiency greater than 60% of that observed when the precursor protein was synthesized in the reticulocyte lysate [25]. These results indicated that post-translational translocation, of mature secretory proteins at least, is not limited to those synthesized in the reticulocyte lysate (which does contain SRP). However, the molecular requirements for targeting and insertion of prepro-melittin into mammalian microsomal membranes appears to differ from that of many of the other proteins discussed above. Like the insertion and processing of of the bacteriophage M13 pro-coat protein [26], prepro-melittin appears to be an exception in that its translocation is not dependent upon the presence of ribosomes, SRP or the SRP receptor molecules. The function of a protein component of the microsomal membrane is required for this translocation process, however, since alkylation of the microsomal membrane proteins by *N*-ethylmaleimide inhibited quantitative sequestration of processed pro-melittin.

ENERGY REQUIREMENTS FOR THE TRANSLOCATION PROCESS

We now know that the mechanism of protein translocation across the ER membrane does not require that protein synthesis and translocation be concurrent in all cases.

Therefore, if energy is required during targeting and translocation, it need not be derived from energy expended during polypeptide elongation. We do not, however, imply that this type of energy cannot play some role in cotranslational translocation. Recent experiments have addressed whether a source of energy is required during the post-translational translocation of proteins across yeast and mammalian microsomes and, if energy is required, from what source is it derived. What is known regarding this question is summarized below.

Post-Translational Translocation Across Yeast Microsomes

Requirement for nucleotides. Depletion of small molecules by gel filtration from a cycloheximide-inhibited translation reaction containing prepro-α-factor resulted in the abolition of post-translational translocation when membranes were subsequently added [20,23]. Translocation could be restored by addition of ATP and an ATP regenerating system [20] or by ATP alone [23], but the reaction could not be restored by addition of nonhydrolyzable analogs of ATP [20,23]. In one report, it was not necessary to desalt the reaction mixture after cycloheximide addition, presumably due to a more rapid exhaustion of the ATP supply in these experiments [21]. However, in order to restore translocation, in the latter report it was sufficient to add ATP and an ATP regenerating system. In all reports to date on this topic, the presence of ATP was not sufficient to restore translocation if an excess of an ATP degrading system was included. Taken together, these results indicate that ATP hydrolysis is required during the post-translational translocation of prepro-α-factor across yeast microsomes. It is a possibility that the ATP is not used directly but instead is used to create some other high-energy compound via enzymes present in the translation extract or the membrane vesicles.

Transmembrane ionic gradient not required. Nucleotide energy could be used to drive protein translocation by generation of a transmembrane ionic potential or a pH gradient. A series of experiments using ionophores and uncouplers showed that the following (alone or in pairwise combinations) had no effect upon the translocation reaction: the proton ionophores CCCP, FCCP, DNP, and SF6847; and the ionophores valinomycin and nigericin [20,21,23]. Therefore it is unlikely that transmembrane ionic gradients play an essential role during post-translational translocation into yeast microsomes in vitro, in contrast to the translocation of proteins across the bacterial plasma membrane [33–35] or mito chondrial envelope [36,37].

Post-Translational Translocation Across Mammalian Microsomes

Requirement for nucleotides. Ribosome-dependent, elongation-independent translocation of the truncated β-lactamase-globin fusion protein was inhibited if the reticulocyte lysate was depleted of small molecules by gel filtration. Translocation could be restored by addition of ATP, GTP and an energy regeneration system [22].

In another study, the ribosome-dependent translocation of the truncated glucose transporter protein was shown to require ATP [38]. Nonhydrolyzable analog of ATP were observed to inhibit translocation. This study indicated that hydrolysis of the terminal phosphoanhydride bond of ATP was essential during elongation-independent translocation. However, GTP was also able to stimulate insertion.

The studies discussed above demonstrated that there is a requirement for the presence of nucleotides during post-translational translocation in the yeast system and elongation-independent translocation in the mammalian system: however, it was not determined whether the nucleotide was required for targeting of the protein to the membrane or during the polypeptide's transmembrane movement. To answer this question, SRP-arrested polysomes synthesizing bovine preprolactin were isolated, and the requirements for the formation of the ribosome membrane junction were investigated [39]. Targeting of the SRP-preprolactin-ribosome complex was shown to require the presence of GTP. In contrast to the study of the insertion of the glucose transporter, in this case the formation of the functional ribosome-membrane junction was also promoted by nonhydrolyzable analogs. Furthermore, after the establishment of the ribosome-membrane junction, translocation of the protein into the lumen of the microsomal vesicle did not require the presence of any nucleotide. The conclusions of this study were that in the ribosome and SRP-dependent translocation reaction the formation of a translocation-competent ribosome-membrane junction requires GTP, but that GTP hydrolysis is not required. This would indicate that a GTP binding protein is involved as an adaptor in coupling the presecretory protein-ribosome complex to the microsomal membrane.

Transmembrane ionic gradient not required. As in the results obtained in the studies of translocation across yeast microsomes, the elongation-independent translocation of the glucose transporter protein across mammalian microsomes was not inhibited by addition of ionophores [38]. Therefore it is not likely that the nucleotide requirement for translocation across mammalian microsomes reflects the need for the generation of an electrochemical potential.

CONCLUSIONS

It has become clear that the mechanism of protein translocation across the endoplasmic reticulum of eukaryotic cells in in vitro reconstituted assay systems does not require that protein translocation be absolutely coupled to protein elongation. There are examples of proteins that can translocate into the ER in vitro in a truly post-translational and ribosome-independent reaction (Fig. 1A). These precursor proteins are likely to exist in several structural forms, which may be interconvertible. Since ATP is required during

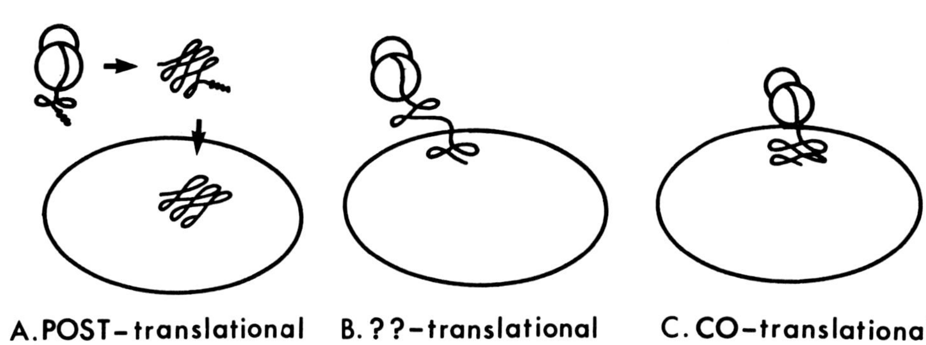

Fig. 1. A) Targeting and translocation of a protein post-translationally occurs after complete elongation of the polypeptide chain and its release from the ribosome. B) Some proteins can be targeted to the ER membrane and begin translocation after a considerable fraction of the polypeptide chain has been synthesized. However, translocation must be initiated before the chain is completely synthesized and released from the ribosome. C) In the cotranslational mode of translocation, targeting is catalyzed by the SRP and SRP receptor system soon after the signal sequence emerges from the ribosome. A ribosome-membrane junction is established, and translocation proceeds, coupled to polypeptide elongation.

translocation of one protein that shows these characteristics, prepro-α-factor, it is tempting to speculate that ATP may be used to influence the structure of the precursor through either a soluble or membrane-associated unfolding factor (Fig. 2) [28,40,41]. The unfolded precursor with signal sequence exposed could then interact directly with a receptor for the signal sequence on the ER membrane that participates in the post-translational, ribosome-independent, and cotranslational processes. Alternatively, ATP could be involved in the assembly of a proteinaceous tunnel in the ER membrane through which the precursor protein passes.

There are proteins however—full-length yeast pre-invertase [47], pre-light chain [9], and preprolactin [7,9], for example—that do require that translocation be initiated during translation before some critical point in elongation. Since translocation can be initiated late during translation for some proteins, or even post-translationally for a few others, the translocation machinery of the ER must either be able to accept and translocate protein domains en bloc, or somehow reduce the folded or partially folded domains to structures compatible with transmembrane movement.

Proteins that are released from the ribosome and then are folded tightly or form oligomeric structures that could mask their signal sequences may be refractile to the attempts of an unfolding factor to act and as a result be translocation-incompetent. In these cases, the kinetic competition between folding and translocation would have more significant effects on translocation competence. Retention on the ribosome of some proteins in this class may serve to inhibit folding of domains deleterious to translocation competence. In addition, the association of a precursor protein with the ribosome appears

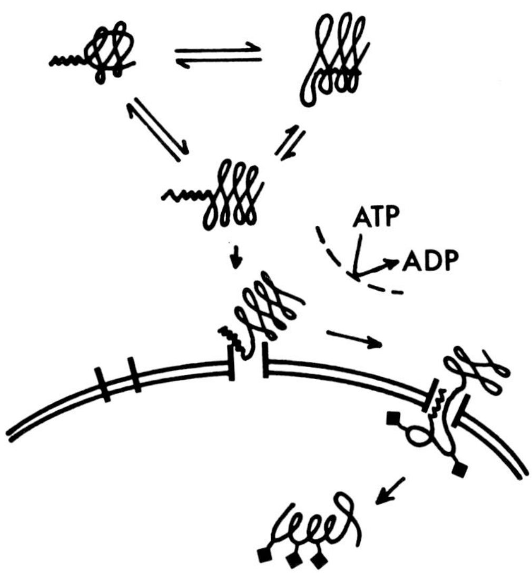

Fig. 2. In the ribosome-independent post-translational translocation reaction, we speculate that prepro-α-factor may exist in different states of interconvertible tertiary structures, in some of which the protein is translocation-incompetent. ATP is required at some point during this mode of translocation. ATP is consumed by either a cytoplasmic or membrane-associated protein. This could happen in the process of rendering a translocation-incompetent substrate competent through a mechanism involving the alteration of the tertiary structure of the protein. Once competent, the protein is translocated through the membrane and glycosylated. The wavy line denotes the signal sequence and the squares denote core oligosaccharides.

to be required for SRP-dependent targeting (see Fig. 1B). Such a requirement would be consistent with SRP's known affinity for the ribosome [42] and the signal sequence [42–44]. This interaction with SRP would in turn greatly increase the efficiency of targeting to the ER membrane, due to the high affinity of SRP for its ER membrane-associated receptor complex [45,46]. Once the precursor protein has been brought into the proximity of additional membrane components involved in transmembrane movement, translocation would then occur.

In cases where protein translocation and translation are tightly coupled, the need for unfolding factors, and thus also ATP, should be abolished (Fig. 1C). When targeting to the ER membrane occurs immediately after emergence of the signal sequence from the ribosome, the primary requirement is the rapid establishment of a functional translocation-competent ribosome-membrane junction. In this type of reaction it has been demonstrated that ATP is indeed not required [39]. Rather, a GTP binding protein is involved in the generation of the functional ribosome-membrane junction and thus seems to serve as an adaptor transiently or permanently between the ribosome and other components of the translocation apparatus of the ER membrane.

Future research will certainly be focused upon purification of the components involved in translocation that require the presence of nucleotides as well as those required during the obligately cotranslational translocation of some proteins across yeast microsomal vesicles. This will allow the purified components functions to be characterized in the reconstituted in vitro assay systems. The ultimate goal of future research in this area will be an understanding of how translocation occurs in vivo, with respect to the relative contributions of the co- and post-translational mechanisms and what roles the identified components of the translocation machinery play in the process.

ACKNOWLEDGMENTS

This work was supported by NIH grant GM-32384 and funds from the Chicago Community Trust/Searle Scholars Program. W. Hansen is a recipient of a NIH Postdoctoral Fellowship (GM-10562).

REFERENCES

1. Blobel G, Dobberstein B: J Cell Biol 67:835, 1975.
2. Walter P, Blobel G: Proc Natl Acad Sci USA 77:7112, 1980.
3. Gilmore R, Walter P, Blobel G: J Cell Biol 95:470, 1982.
4. Meyer DI, Krause E, Dobberstein B: Nature 297:647, 1986.
5. Walter P, Gilmore R, Blobel G: Cell 38:5, 1984.
6. Hortsch M, Meyer, DI: Int Rev Cytol 102:215, 1986.
7. Siegel V, Walter P: J Cell Biol 100:1913, 1985.
8. Siegel V, Walter P: Nature 320:81, 1986.
9. Ainger KJ, Meyer DI: EMBO J 5:951, 1986.
10. Lipp J, Dobberstein B, Haeuptle, MT: J Biol Chem 262:1680, 1987.
11. Rothman JE, Lodish H: Nature 269:775, 1977.
12. Braell WA, Lodish H: Cell 28:23, 1982.
13. Rapoport TA, Heinreich R, Walter P, Schulmeister T: J Mol Biol 195:621, 1987.
14. Randall LL, Hardy SJS: "Modern Cell Biology." New York: Allen R. Liss Inc., 1982.
15. Wickner W, Lodish H: Science 230:400, 1985.
16. Hay R, Bohni P, Gasser S: Biochim Biophys Acta 779:65, 1984.
17. Harmey MA, Neupert W: "The Enzymes of Biological Membranes," Vol 4. New York: Plenum, 1985.

18. Mullet JE, Grossman AR, Chua NH: "Cold Spring Harbor Symposia on Quantitative Biology," XLVI. New York: Cold Spring Harbor, 1982.
19. Mueckler M, Caruso C, Baldwin S, Panico M, Blench I, Morris H, Allard W, Lienhard G, Lodish H: Science 229:941, 1985.
20. Hansen W, Garcia PD, Walter P: Cell 45:397, 1986.
21. Rothblatt J, Meyer DI: EMBO J 44:1031, 1986.
22. Perara E, Rothman RE, Lingappa VR: Science 232:348, 1986.
23. Waters MG, Blobel G: J Cell Biol 102:1543.
24. Caulfield MP, Duong LT, Rosenblatt M: J Biol Chem 261:10953, 1986.
25. Zimmerman R, Mollay C: J Biol Chem 261:12889, 1986.
26. Watts C, Wickner W, Zimmerman R: Proc Natl Acad Sci USA 80:2809, 1983.
27. Hortsch M, Avossa D, Meyer DI: J Cell Biol 103:241, 1986.
28. Waters MG, Chirico WJ, Blobel G: J Cell Biol 103:2629, 1986.
29. Stevens T, Esmon B, Schekman R: Cell 30:439, 1982.
30. Eilers M, Schatz G: Nature 322:228, 1986.
31. Randall LL, Hardy SJS: Cell 46:921, 1986.
32. Mueckler MK, Lodish H: Cell 44:629, 1986.
33. Date T, Zwizinski C, Ludmerer S, Wickner W: Proc Natl Acad Sci USA 77:827, 1980.
34. Bakker E, Randall LL: EMBO J 3:895, 1984.
35. Geller BL, Movva NR, Wickner W: Proc Natl Acad Sci USA 83:4219, 1986.
36. Reid GA, Yonetoni T, Schatz G: J Biol Chem 257:10368, 1982.
37. Pfanner N, Neupert W: EMBO J 4:2819, 1985.
38. Mueckler MK, Lodish H: Nature 322:549, 1986.
39. Connolly T, Gilmore R: J Cell Biol 103:2253, 1986.
40. Rothman JE, Kornberg RD: Nature 322:209, 1986.
41. Zimmerman R, Meyer DI: TIBS 11:512, 1986.
42. Walter P, Ibrahimi I, Blobel G: J Cell Biol 91:545, 1981.
43. Kurzchalia TZ, Wiedmann M, Girshovich AS, Bochkareva ES, Bielka H, Rapaport TA: Nature 320:634, 1986.
44. Krieg U, Walter P, Johnson A: Proc Natl Acad Sci USA 83:8604, 1986.
45. Lauffer L, Garcia PD, Harkins RN, Coussens L, Ullrich A, Walter P: Nature 318:334, 1985.
46. Tajima S, Lauffer L, Rath VL, Walter P: J Cell Biol 103:1167, 1986.
47. Hansen W, Walter P: J Cell Biol (in press), 1988.

Molecular Biology of Intracellular Protein
Sorting and Organelle Assembly 215–219 (1988)

Proteins Containing Covalently Bound Lipid

Bartholomew M. Sefton

Molecular Biology and Virology Laboratory, Salk Institute, San Diego, California 92138

Proteins in eukaryotic cells are subject to three forms of modification with lipid. Some proteins found in the cytoplasm contain myristic acid at their amino termini. A second group of proteins contain palmitic acid linked to cysteine within domains that interact with membranes. The third group is anchored to the cell surface by an exotic phospholipid tail. The mechanisms by which these several lipids are transferred to proteins are distinctly different.

Key words: myristic acid, palmitic acid, phosphatidylinositol tail

Eukaryotic proteins undergo three forms of covalent modification with lipid: myristylation (Table I), palmitylation (Table II), and replacement of a carboxy-terminal domain with a phospholipid tail (Table III). These are remarkably different processes.

MYRISTYLATION

Myristic acid, the 14-carbon-saturated fatty acid, is always found linked through an amide bond to the alpha amino group of amino-terminal glycine [1,2]. No other fatty acid has been found to modify the amino termini of eukaryotic proteins. Additionally, only amino terminal glycine undergoes myristylation. Point mutations that replace amino terminal glycine with alanine abolish myristylation [3,4]. The enzyme that attaches lipid to amino termini obviously has great specificity for both fatty acid and amino acid substrate.

Myristylation occurs very early during the lifetime of a protein. Nascent polypeptide chains may in fact be the substrates of the myristyl transferase [5]. The lipid donor is most probably myristyl CoA [6]. Amino terminal myristate does not undergo turnover [5].

The function of protein myristylation is understood in a number of cases. The transforming protein of Rous sarcoma virus, $p60^{v-src}$, is normally found on the cytoplasmic face of the plasma membrane of infected cells. Point mutations replacing the myristylated glycine in this protein have no effect on its protein kinase activity, yet render it both soluble and unable to transform cells [3,7,8]. Similarly the internal structural

Received July 8, 1987.

TABLE I. Myristylated Proteins

Protein	Reference
p60^{src}	2
Catalytic subunit of cAMP-dependent protein kinase	1
p15^{gag} of Moloney murine leukemia virus	31
P120$^{gag\text{-}abl}$ transforming protein	32
p56^{lck}	33,38
VP2 of papovaviruses	9
VP4 of picronaviruses	10

protein of murine leukemia virus, p15^{gag}, normally binds tightly to cellular membranes during virion formation. Point mutations replacing the myristylated amino-terminal glycine of p15^{gag} yield a soluble protein [4]. It is clear therefore that the myristylation of these two proteins is essential for their stable association with cellular membranes. In these cases, the ability of a protein to bind to a membrane is apparently acquired through post-translational modification, rather than encoded in the gene.

Myristylation apparently also has other functons. It has been known for some time that not all myristylated proteins are bound to cellular membranes [5]. Indeed, it has been shown recently that the amino termini of two structural proteins of viruses whose virions do not contain lipid bilayers, the VP2 protein of the papovaviruses [9] and the VP4 protein of picornaviruses such as poliovirus [10], are myristylated. The structure of the poliovirus particle is in fact understood in atomic detail. The myristate moiety in VP4 is buried in a region of five-fold symmetry within the water-soluble virion [10]. Apparently, it forms a hydrophobic center during the assembly of the viral particle.

Picornaviruses undergo a conformational change when they bind to receptors on the surface of susceptible cells. VP4 is in fact lost from the viral particle if the adsorbed virus becomes dissociated from the cell surface [11]. The details of this conformational change are not understood. The myristyl moiety in VP4 could become exposed as a result of virus attachment and play a role in virus penetration. It is obviously important now to learn whether a mutation preventing myristylation of VP4 affects either virus assembly, structure, or infectivity.

PALMITYLATION

Palmitic acid, the 16-carbon-saturated fatty acid, is almost always found linked through a thioester bond to cysteine [12]. The palmitic acid present in rhodopsin is linked through an ester bond to threonine, but it may not be incorporated through an enzymatic process [13]. Early reports that palmitic acid is linked to the G glycoprotein of vesicular

TABLE II. Palmitylated Proteins

Protein	Reference
p21^{ras}	37,38
Vesicular stomatitis virus G glycoprotein	35
Alphavirus E2 glycoprotein	36
Mammalian transferrin receptor	39
Apolipoprotein A	15
Ankyrin	18

TABLE III. Proteins Containing a Phospholipid Tail

Protein	Reference
Thy-1 glycoprotein	23,24
Variant surface glycoprotein of trypanosomes	25
Acetylcholine esterase	40
5′-nucleotidase	41
Alkaline phosphatase	42
NCAM-120	43
TAP glycoprotein	44

stomatitis virus through an ester bond to serine were probably incorrect. A cysteine is essential for palmitylation. Mutations that convert palmitylated cysteines to serine abolish acylation [12,14].

Sites of palmitylation are usually present in a domain of a membrane-bound protein that resides at or near the cytoplasmic face of membranes. This is not, however, universal. The secreted protein apolipoprotein A1 is palmitylated [15]. Where and when acylation of apolipoprotein A1 occurs is not known.

Unlike myristylation, palmitylation is clearly a post-translational event. Viral surface glycoproteins become palmitylated approximately 20 minutes after their release from the polysome [16]. This occurs as these proteins pass through the Golgi apparatus, during their transport to the cell surface. The lipid donor is probably palmityl CoA [17]. The fact that palmitylation can occur post-translationally is also apparent from the findings that ankyrin can be labeled with radioactive palmitic acid in mature red cells that are no longer synthesizing the protein [18] and that the transferrin receptor can be palmitylated at least 24 hours after its synthesis [20].

The palmitylation of proteins long after their synthesis could reflect either slow, incomplete palmitylation of the protein, or the turn-over of palmitate in protein. Recent pulse-chase experiments suggest strongly that the palmityl moieties in both $p21^{N-ras}$ [21] and ankyrin undergo fairly rapid turn-over [18]. This cycling may play a role in the function of these proteins.

The function of protein palmitylation is not understood. Mutations that prevent the palmitylation of the mammalian transferrin receptor and the G glycoprotein of vesicular stomatitis virus have no effect on the transport of the proteins to the cell surface [12,22]. Indeed, natural nonpalmitylated variants of the G glycoprotein exist [12]. In contrast, point mutations replacing the palmitylated cysteine in $p21^{ras}$ abolish acylation and render the protein both unable to bind to cellular membranes, and unable to induce cellular transformation [14]. The direct demonstration that the absence of palmitic acid *induces* these functional changes, rather than *reflecting* them is, however, so far lacking.

ATTACHMENT OF A COMPLEX PHOSPHOLIPID TAIL

Perhaps most unexpected is the finding that a remarkably large number of cell surface glycoproteins are bound to the cell surface not by a transmembrane protein domain but rather by a phospholipid tail (Table III). The Thy-1 glycoprotein and the variant surface glycoprotein of trypanosomes are the prototypes of this class of protein [23–25].

Both proteins contain a complex, glycosylated form of phosphatidylinositol linked through an amide bond between their carboxy-terminal carboxyl group and ethanolamine, one of the constituents of the novel phospholipid. Unlike palmitate and myristate, which are always linked to cysteine and glycine respectively, phospholipid tails have been found linked to a number of different carboxy-terminal amino acids.

The lipid tail is not in fact attached to the true carboxy terminus of the protein. Proteolytic processing precedes the addition of the tail. Both the Thy-1 glycoprotein and the variant surface glycoprotein are synthesized as precursors that contain a hydrophobic carboxy-terminal polypeptide domain [26] This portion of the protein is cleaved proteolytically within a minute of the release of the protein from the polysome and replaced with the exotic lipid [27].

The exact composition and structure of phospholipid tails are not understood. They contain at a minimum phosphatidylinositol, ethanolamine, galactose, glucosamine, and galactosamine. The donor of the tail has not been identified. It is likely that the glycosylated lipid is preassembled and transferred in one piece, rather than being assembled sequentially on the polypeptide chain (see below).

Two reasonable questions are: 1) why are proteins anchored on the cell surface by a phospholipid tail; and 2) why do phospholipid tails have such a bizarre structure? One possible property of a lipid tail is that it allows the protein to diffuse rapidly in the plane of the plama membrane bilayer, without interference from the cytoskeleton. Indeed, there is some evidence consistent with this notion [28].

The fact that phospholipases can hydrolyze the bond between the protein and the lipid tail anchoring it to the cell surface may also be of significance. This susceptibility to phospholipases provides a novel means through which a protein can be modulated or removed from thc cell surface. In addition, cleavage of the lipid tail by a phospholipase C will release a physiologically significant lipid, diacylglycerol, the natural activator of protein kinase C. The cleavage of the lipid tail of a protein is therefore potentially a means of activating protein kinase C and coupling the release of the protein to intracellular events.

In this context, the recent observation that the binding of insulin to its receptor leads to the generation of novel, water-soluble forms of glycosylated inositol phosphate [29] is provocative. These glycosylated inositol phosphates closely resemble those found in the phospholipid tails of surface proteins. There are many reasons to think that inositol phosphates can function as second messengers for factors that bind to cell surface receptors [30], and the same may be true of inositol phosphates that contain carbohydrate. One interpretation of the effect observed with insulin is that the binding of insulin to its receptor activates a phospholipase that hydrolyzes the donor of the complex lip tail found in glycoproteins, releasing diacylglycerol and a glycosylated inositol phosphate. It is possible that proteins with phospholipid tails can also serve as a source of inositol-containing second messengers and that their enzymatic hydrolysis might repesent a new form of transmembrane signal transmission.

Given the rapidly increasing number of proteins known to contain lipid, there seems little doubt that this is an important and unexpectedly common form of protein modification. Many new insights into the biochemistry and function of lipid modification can be anticipated in the near future.

REFERENCES

1. Carr, SA, Biemann K, Shoji S, Parmelee DC, Titani K: Proc Natl Acad Sci USA 79:6128, 1982.
2. Schultz AM, Henderson LE, Oroszlan S, Garber EA, Hanafusa H: Science 227:427, 1985.

3. Kamps MP, Buss JE, Sefton BM: Proc Natl Acad Sci USA 82:4625, 1985.
4. Rein A, McClure MR, Rice NR, Luftig RB, Schultz AM: Proc Natl Acad Sci USA 83:7246, 1986.
5. Buss JE, Kamps MP, Sefton BM: Mol Cell Biol 4:2697, 1984.
6. Towler DA, Eubanks SR, Towery DS, Adams SP, Glaser L: J Biol Chem 262:1030, 1987.
7. Buss JE, Kamps MP, Gould K, Sefton BM: J Virol 4:2696, 1986.
8. Kamps MP, Buss JE, Sefton BM: Cell 45:105, 1986.
9. Streuli CH, Griffin BE: Nature 326:619, 1987.
10. Chow M, Newman JFE, Filman D, Hogle JM, Rowlands, DJ, Brown F: Nature 327:482, 1987.
11. Lonberg-Holm K, Gosser LB, Kauer JC: J Gen Virol 27:329, 1975.
12. Rose JK, Adams GA, Gallione CJ: Proc Natl Acad Sci USA 81:2050, 1984.
13. O'Brien PJ, St Jules RS, Reddy TS, Bazan NG, Zatz M: J Biol Chem 262:5210, 1987.
14. Willumsen BM, Norris K, Papageorge AG, Hubbert NL, Lowy DR: EMBO J 3:2581, 1984.
15. Hoeg JM, Meng MS, Ronan R, Fairwell T, Brewer HB Jr: J Biol Chem 261:3911, 1986.
16. Schmidt MFG, Schlesinger MJ: J Biol Chem 255:3334, 1980.
17. Berger M, Schmidt MFG: J Biol Chem 259:7245, 1984.
18. Staufenbiel M, Lazarides E: Proc Natl Acad Sci USA 83:318, 1986.
19. Omary MB, Trowbridge IS: J Biol Chem 256:12888, 1981.
20. Magee AI, Guitierrez L, McKay I, Hall A, Marshal C: J Cell Biochem 11A:261, 1987.
21. Jing S, Trowbridge IS: EMBO J 6:327, 1987.
22. Low MG, Kincade PW: Science 318:62, 1985.
23. Tse AGD, Barclay AN, Watts A, Williams AF: Science 230:1003, 1985.
24. Ferguson MAJ, Low MG, Cross GAM: J Biol Chem 260:14547, 1985.
25. Seki T, Moriuchi T, Chang H-C, Denome R, Silver J: Nature 313:485, 1985.
26. Bangs JD, Gereld D, Krakow JL, Hart GW, Englund PT: Proc Natl Acad Sci USA 82:3207, 1985.
27. Ishihara A, Hou Y, Jacobson K: Proc Natl Acad Sci USA 84:1290, 1987.
28. Saltiel AR, Cuatrecasas P: Proc Natl Acad Sci USA 83:5793, 1986.
29. Majerus PW, Connolly TM, Deckmyn H, Ross TS, Bross TE, Ishii H, Bansal VS, Wilson DB: Science 234:1519, 1986.
30. Henderson LE, Krutzsch HC, Oroszlan S: Proc Natl Acad Sci USA 80:339, 1983.
31. Schultz A, Oroszlan S: Virology 133:431, 1984.
32. Voronova AF, Buss JE, Patschinsky T, Hunter T, Sefton BM: Mol Cell Biol 4:2705, 1985.
33. Marchildon GA, Casnellie JE, Walsh KA, Krebs EG: Proc Natl Acad Sci USA 81:7679, 1984.
34. Schmidt MFG, Schlesinger MJ: Cell 17:813, 1979.
35. Schmidt MFG, Bracha M, Schlesinger MJ: Proc Natl Acad Sci USA 76:1687, 1979.
36. Sefton BM, Trowbridge IS, Cooper JA, Scolnick EM: Cell 31:465, 1982.
37. Chen Z-Q, Ulsh LS, DuBois G, Shih TY: J Virol 56:607, 1985.
38. Omary MB, Trowbridge IS: J Biol Chem 256:4715, 1981.
39. Futerman AH, Fiorini R-M, Roth E, Low MG Silman I: Biochem J 226:369, 1985.
40. Low MG, Finean JB: Biochim Biophys Acta 508:565, 1978.
41. Low MG, Zilversmit DB: Biochemistry 19:3913, 1980.
42. He H-T, Barbet J, Chaix J-C, Goridis C: EMBO J 5:2489, 1986.
43. Reiser H, Oettgen H, Yeh ETH, Terhorst C, Low MG, Benacerraf B, Rock KL: Cell 47:365, 1986.

Index